QUANTUM REALITY
THE STORY OF QUANTUM PHYSICS

QUANTUM REALITY
THE STORY OF QUANTUM PHYSICS

ROBERT RANKIN

RANKIN PUBLISHERS
BRISBANE

Published by
Rankin Publishers
PO Box 350
Kenmore Qld 4069
AUSTRALIA

www.rankin.com.au
info@rankin.com.au

Quantum Reality
The story of quantum physics

First Edition 2020

ISBN 978 0 9874938 7 3

Prepress by Rankin Publishers, Brisbane
AUSTRALIA

Cover design by Ben Rankin

CONTENTS

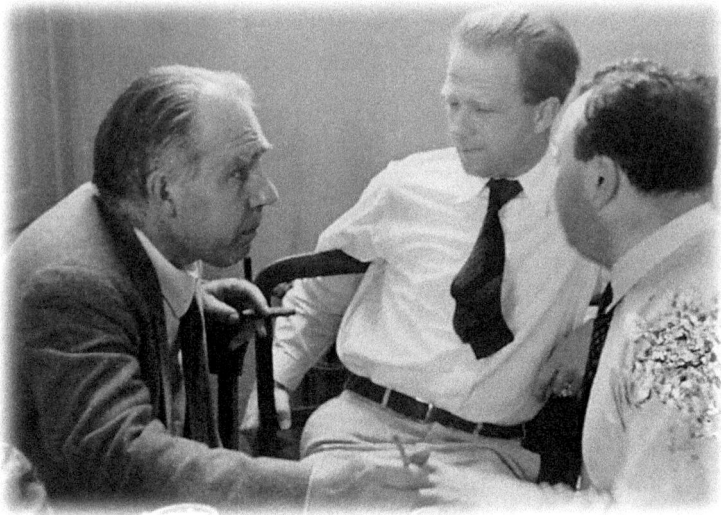

Niels Bohr, Werner Heisenberg and Wolfgang Pauli at the Niels Bohr Institute at the Copenhagen Conference, 1936. Between 1913 and 1926, well before this photograph was taken, these physicists and others had put together the basic structure of the quantum theory of nature. (Courtesy of the Niels Bohr Archive, Copenhagen, Denmark)

CHAPTER 1

INTRODUCTION

By their very nature, scientific concepts often build one on top of the other. To understand a new idea, it is usually necessary to comprehend the concepts which led up to it. This is often what makes science so hard to follow. Scientific concepts are not isolated facts such as knowing the dates at which historic events occurred or knowing the scientific name of a plant species. Scientific theories develop from a long line of deductions using simpler concepts and, to fully appreciate an all-encompassing theory, means the subordinate ideas must be firstly understood.

This book is written in such a deductive style, starting with the basic ideas and building on them. So, it is intended for the reader to start at the first page and follow the argument in a linear fashion. If, at some point, you get stuck and cannot follow the argument, it is advisable to re-read the section. Having said that, it is still possible to jump forward. Each chapter of the book discusses a major part of the overall topic and it is possible to read and understand the chapters in isolation, but the fullest understanding will come from reading the complete text in order.

A lot of the concepts discussed are quite difficult to understand. You may need to re-read some paragraphs several times to grasp their significance but if after doing so you are still confused, move on to the next section and come back to the obstacle again at a later time.

The appendices at the end of the book explain issues mentioned in the text in a more mathematical way. They will be of interest to those with a basic understanding of mathematics. As well, I have included several equations in the main text. Again, for those who know some algebra, these equations will be of interest. However, even without mathematical knowledge there is a lot to be gained by reading the book and simply ignoring the mathematics. Skip the equations and concentrate on the verbal descriptions. By the end of the book you will still have gained a great deal of appreciation for the issues involved in the discipline of quantum physics.

Normal mathematical conventions are used in presenting the equations. Square roots are shown with the usual square root sign. When there is a need to show a divided by b in the text, this is written as a/b. Equations do not usually incorporate multiplication signs, so a multiplied by b, or $a \times b$, is usually written as ab. Also, $a \times a$, or a squared is written a^2.

Finally, I would like to thank Michael Kewming from the School of Mathematics and Physics at the University of Queensland, Australia, for his thoughtful scientific advice, and the American Institute of Physics Emilio Segre Visual Archives and the Niels Bohr Archive, Denmark, for permission to publish the historical photographs.

CHAPTER 2

THE CRUX OF THE ISSUE

When the phone rang at four in the morning, Richard Feynman knew what to expect. It was October 21, 1965. Nobel Prizes do not come easily or quickly to theoretical physicists but they do come early in the morning for those living in the USA. It had been nearly two decades since Feynman and two other researchers had unravelled the secrets of quantum electrodynamics. Theoretical ideas need to be confirmed before the scientific community tends to accept them and now after two decades enough confirmation had been received. Together with Julian Schwinger and Shinichiro Tomonaga, Feynman was at last being recognized for the work he had done in the late 1940s in this most esoteric but highly relevant field. But I am getting well ahead in the full story of the development of quantum physics. Its birth was at the beginning of the twentieth century and Richard Feynman had not yet been born. He was but one researcher in a long line who contributed to this most arcane discipline.

It is no surprise that Albert Einstein was involved and was one of a hand full of pioneers. Everyone has heard of Einstein's theory of relativity even if they do not understand it. Einstein produced his first theory of relativity, the special theory, in 1905 and the general theory in 1916. These were his greatest discoveries, so it comes as some surprise to learn that he did not receive his Nobel Prize for either of them although both were certainly worthy of it. Such was his genius that to whatever he turned his hand great discoveries ensued. Einstein received the Nobel Prize in 1921 in a field which

9

at the time had no name but later would become known as quantum physics. Specifically, his research related to the photoelectric effect—a phenomenon which surprisingly could only be successfully explained using the new quantum concepts.

Together with a small group of contemporaries, Einstein can certainly be classed as a quantum pioneer and initially a strong supporter of its often-radical ideas. Yet within a few short years he was denouncing the entire discipline. In fact, for the rest of his life he could not reconcile what quantum physics had to say about the world with his ideas of how it should be.

Another pioneer of whom we will hear quite a lot later is the Austrian physicist Erwin Schrödinger. When he realized what quantum physics implied, he was quoted as saying that had he known he and his contemporaries were not going to eliminate this *damned quantum jumping* (a term that will be explained later), he never would have involved himself in this business in the first place. (1)

These older pioneers had let the quantum genie out of the bottle and there was no way of putting it back. In fact, nobody really wanted it put back because, as time went by, the quantum theory became a very valuable and practical theory which explained chemical and nuclear reactions and was used to develop a whole host of new inventions including television, transistors and lasers. It was the theory's far-reaching and philosophical implications which some researchers found distasteful. Others, mostly of a younger generation, found them quite acceptable and without a second thought forged ahead with the theory. Because of its widespread use and practical application, quantum physics today is fully accepted as are mostly its strange implications.

Like so many scientific ideas when first proposed, quantum physics contradicted established ideas explaining how nature worked. In time, the implications of most theories become accepted as quite obvious and ordinary but with the quantum theory this never happened. The implications were too profound or bizarre to accept and scientists continually thought there must be more. They assumed the theory as it stood was only half the story—that the missing links would one day be found to make it more acceptable.

Today, it appears we do have a full theory and the theory stands as is. It must be accepted that the world behaves according to these strange

10

implications, ones that turned the classical view of the physical world on its head. Shinichiro Tomonaga once said *Why isn't nature clearer and more directly comprehensible?* No one really knows.

How, Not Why

Nobody understands why nature is as it is although scientists often loosely use the word *understand* to imply they comprehend such-and-such a theory. What is really meant by scientific understanding is an understanding of *how* nature works. Science can explain how things work, often to a very accurate degree. As such, science can make predictions about systems. For example, knowing how much water is in a kettle and also knowing the rate at which heat is being added to this water, it can easily be worked out exactly when the kettle will boil. That is all about knowing the how of nature. Why a certain quantity of water requires a certain amount of heat to make it boil can never be answered. Simply, that is the way nature is. Science cannot answer why nature is this way, only how it operates.

Science is similar to an instruction manual for driving and maintaining a motor car. The manual can explain how to drive and fix the car but it does not explain why the car is designed the way that it is. It would, however, be quite easy to write a manual explaining why cars are so designed but it is not normally the role of car manuals. Science has, so far, done a pretty good job of writing the manual explaining how nature operates but it certainly has not even begun to write the first page of the manual explaining why nature behaves as it does. The fine distinction between *how* and *why* needs to be explained here if you are to accept what quantum physics has to say about nature.

Nobody understands quantum physics. Perhaps more precisely, nobody understands why nature behaves according to the rules of quantum physics. It just does and this brings us to another issue in science.

Nature is Absurd

Many people find the theories of science difficult to comprehend. There are several reasons for this. Firstly, they are often difficult because they are often quite complex and require a lot of mental effort to grasp. Secondly, they are often unintuitive. Time and time again nature is seen to operate in strange ways—ways very different to what is expected or would be considered obvious.

For example, many years ago, our eyes were thought to operate by sending out some strange rays which interacted with the environment and then sending back messages to our brain so that we knew what objects looked like. This early theory of the eye seems absurd now with our current level of scientific knowledge but as a child I can remember thinking this way. The reason for this incorrect view probably stems from our concept of touch. To investigate an object in front of us we extend our arm and touch it with our fingers, so gaining some idea of the form of the object. We are taking an active role in the investigation by extending our arm to the object. Logically, it might be argued our eyes do the same by sending out mysterious beams to interrogate an object. As it turns out, our eyes are more passive than that and simply receive light rays from the object. Our brain then gets active and interrogates the image on the retina of the eye. The reason that this correct theory was not thought of first is because it did not relate to any ideas which had gone before. It was a new concept and formulating a new concept takes great effort and often a giant intellectual leap.

Another reason scientific ideas may take time to be understood or accepted involves a value judgement. The new ideas may seem so absurd that you cannot accept them. You do not like what the theory is telling you and you conclude that nature cannot surely be so badly designed. There must be a better theory.

Theories are often termed scientific models. These are not models in the sense of something smaller than the real thing as in a model railway. Rather, the concept of a scientific model has developed because scientists realize their theories are really only analogies or metaphors that describe how nature operates in terms of something already known. We say an apple drops from a tree because gravity pulls it towards the earth like a magnet. We see magnets attracting metal objects and this makes it easier to think that solid earth may be able to also attract things. So, a new theory is built up around a familiar idea and extended and adapted where it seems necessary. With this approach we accept that we never really have the last say. Models are always open to criticism and further adjustment as new information comes to light. Sometimes models are thrown out completely when it is realized they lead up a blind alley and are useless in describing new ideas or discoveries. Such is the process of science.

Explaining a Complex Theory

There are many ways to explain a complex theory. One traditional method is to follow the path of discovery step-by-step as researchers gradually unravel the mysteries. This historical method has the advantage of highlighting the issues as they logically arise and explaining how each in turn is dealt with. Putting a theory in its historical context often assists in its understanding and this method is mostly used here.

There is another way. With the advantage of hindsight—that is, after the theory has been fully developed, it often becomes obvious that the historical path taken by the pioneering researchers may not in fact be the best way to explain the theory. Many blind and unhelpful alleys may have been followed and what should have been obvious paths ignored. With hindsight, the theory may better be presented out of historical sequence. As well, as time passes, the historical context becomes less relevant, so explanations tend to follow a logical development independent of history. This will occur more and more with the theory of quantum physics.

There is an unsolved mystery in science as to whether scientific ideas are discovered in the only possible order or whether, if mankind had its time over, the sequence of discoveries would be in a different order. Quite possibly many would be but, generally, the path of discovery would probably be similar to the one that did evolve. Certainly, some theories would have had to come first because others are built on them. For instance, it is hard to imagine the theory of relativity proposed by Albert Einstein (1879 – 1955) preceding the work on the motion of bodies put forward by Sir Isaac Newton (1642 – 1727) because Einstein's theory is an extension to Newton's. It would have taken a massive intellect to make the jump to Einstein's ideas in one move without first proposing Newton's theory.

So it is with quantum physics. Superficially, the world does not seem to operate according to these strange quantum rules. There would have been no reason for early researchers to propose a quantum theory because their knowledge of the universe was not that deep. Only later, as all the pieces of a physical theory of nature were painstakingly put together did it become obvious that some of the more recent discoveries did not fit at all well with the contemporary theory. Adjustments were obviously needed but mere tinkering with this current theory was of no help. A radical re-think was required and so the quantum theory was born.

It is quite clear that the classical theories of physics had to pre-date the quantum theory. Classical theory grew out of everyday experience—how we think the world works. This model served society well and still does in many scientific fields. Even Einstein's theory of relativity belongs to this classical era but, surprisingly, almost at the same time he was presenting this theory of relativity, Einstein was also, almost unknowingly, sowing the seeds of quantum physics with his investigations of light and the photoelectric effect. But more on that later.

A good place to start with an explanation of quantum ideas is with the nature of light and indeed it was this study of light which ultimately began the quantum era. Light had been studied for many centuries and, even in Newton's day, was considered an odd commodity.

The Nature of Light

Today the generic term *light* refers generally to the entire electromagnetic spectrum—the whole gamut of electromagnetic waves which permeate our lives and include radio and television waves, infra-red rays, the visible light rays (red, orange, yellow, green, blue, indigo and violet), ultra-violet rays, x-rays and gamma rays. All these waves are the same commodity. Each group simply covers a different frequency range.

By frequency, scientists mean the number of waves passing a point per second—exactly the same idea as counting the number of ocean waves which might break onto a beach every second although, in that example, the number of ocean waves per minute, rather than per second, might be a better measure. Compared to light, ocean waves have very low frequency.

Another associated term is wavelength. This is easier to understand than frequency and is simply the distance between wave crests. For surfing waves near the beach this could be about twenty to fifty metres and for ripples on a pond much less than one metre. The amplitude of a wave is the height the crest rises above the flat ocean. The height from a crest to a trough is then equal to twice the wave's amplitude. Figure 1 illustrates these terms.

So the only difference between visible red light and X-rays, for example, is the wave's frequency. They are certainly not different commodities. The Scottish physicist, James Maxwell (1831 – 1879), cleared that up in the 1860s when he formulated a theory that proposed light to be a combination of oscillating electric and magnetic fields. Hence, in scientific terms, the name for a light wave became an electromagnetic wave. In the early 1890s, the German physicist Heinrich Hertz (1857 – 1894) confirmed this was indeed the case by experiment. The actual discovery of very high frequency

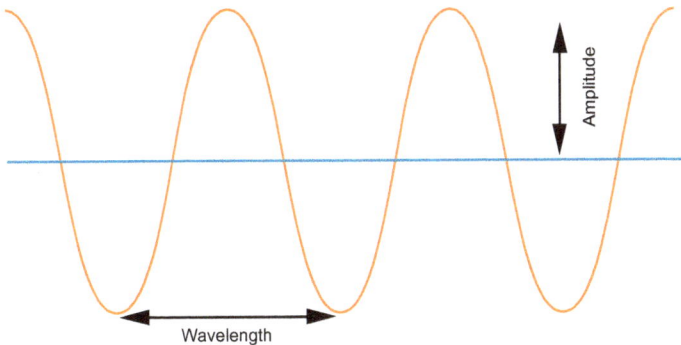

FIGURE 1: Wave terminology. The vertical axis displays the size, strength or amplitude of the wave. The horizontal axis can display either distance or time. In this case, it is distance so the length between successive crests or troughs is the wavelength.

electromagnetic waves, such as x-rays, was, of course, still a long way off but the groundwork had now been laid.

Today we have no trouble accepting light to be a wave-like phenomenon, but this was not always the case. In the 1600s, Newton had done a very good job of describing how bodies moved under the influence of forces and gravity. He developed his theories of motion and his equations are still used today in many applications. It is not surprising then that when Newton began studying the behaviour of light he did so by first assuming that light was made up of tiny particles which he called corpuscles. This theory had a lot going for it. Firstly, objects such as a ball (which obeyed Newton's laws of motion), when thrown tended to move in roughly a straight line until it hit a wall and bounced off. Light from the sun seemed to do the same. We cast a shadow because our body intercepts light particles travelling in straight lines from the sun, so preventing them from striking the ground. Light also bounces off a mirror in a way similar to a ball bouncing off a wall.

Newton's corpuscular theory could not however satisfactorily explain refraction—the bending of light through denser-than-air substances such as glass or water, giving us such practical items as

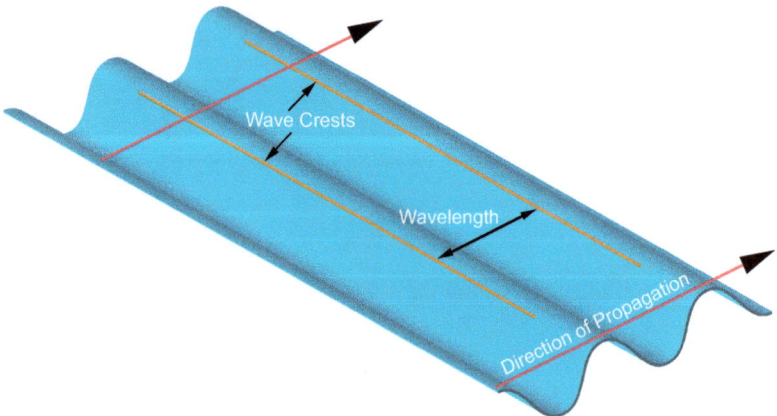

FIGURE 2: The concepts of direction of propagation, wave crests and wavelength as depicted using water waves.

lenses in cameras and spectacles. Newton's theory of light required light to travel more quickly through denser objects, something today we know is wrong. Light travels more slowly through glass than it does through air.

Although this was not known in Newton's time, the Dutch mathematician and physicist Christiaan Huygens (1629 – 1695), working at the same time as Newton, proposed a radical wave theory for light which described everything that Newton's particle theory did but also required light to travel more slowly through glass or water than it does through air. In this regard, Huygens theory was correct but this was not realized at the time because no one had developed a means of measuring the speed of light. Both theories co-existed as an explanation of the behaviour of light with Newton's theory dominating for a century probably because of his scientific status.

It took a simple little experiment by today's standards to finally vindicate the wave theory. English physicist Thomas Young (1773 – 1829) and his French counterpart Augustin Fresnel (1788 – 1827) working independently in the early 1800s showed that light really did behave like a wave. The experiment is the famous double-slit experiment which every high school or early university physics student has performed now for almost two centuries. It once and for all explains light as a wave-like phenomenon by showing the effects of two waves interfering due to diffraction.

Now interference and diffraction may well be two very complex scientific terms, but they can be fairly easily explained using more familiar types of waves such as ocean waves which behave in exactly the same way as light waves.

The Double Slit Experiment

The wave theory of light took a long time to be recognised because the particle theory explained the properties of light so well—that it travelled in straight lines so producing shadows and that it bounced off walls like a ball. Flaws in the argument were never seen because the details were not looked at closely.

As Huygens had shown, the wave theory could also be used to demonstrate these properties but at the time no one really listened to him. In wave theory, three basic concepts are the direction of

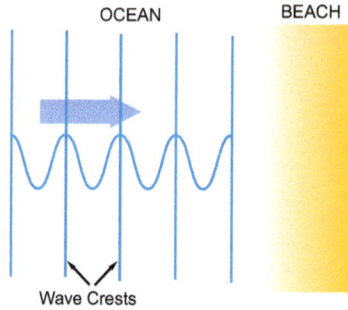

FIGURE 3a: Ocean wave crests travel in straight parallel lines towards a soft sandy beach in the direction of the large blue arrow, dissipating their energy into the sand.

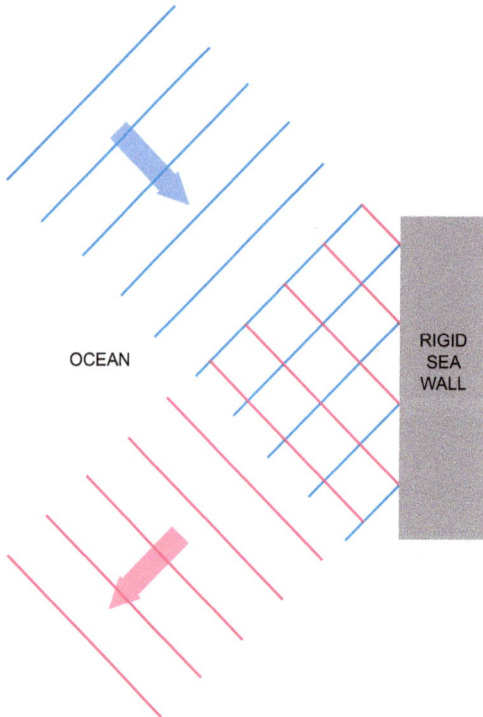

FIGURE 3b: The same travelling ocean waves as in Figure 3a above (blue wave crests) reflect or bounce off a hard rigid sea wall (pink wave crests).

18

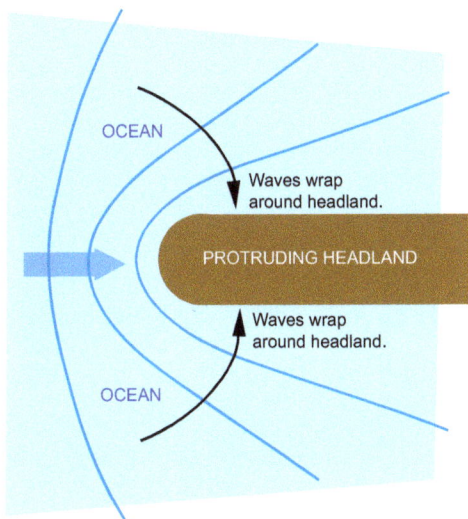

FIGURE 3c: Incoming parallel ocean wave crests wrap around a protruding headland.

propagation, the idea of wave crests and troughs, and the idea of wavelength—the distance between two adjacent crests or troughs. These are shown in Figure 2.

As Figures 3a and 3b show, ocean waves travel across the open water in straight lines until they crash onto a beach or hit a sea wall. When they do strike an abrupt rocky wall, they are seen to reflect back just like a ball and even demonstrate refraction, or bending, when they wrap around a headland and break onto the beaches on either side of a headland as in Figure 3c. Further, ocean waves will travel through a large seaway opening to a harbour almost unaltered and will therefore require a baffle wall on which to break. See Figure 3d.

Now here is the crunch. If the seaway is made very small, much smaller than the wavelength of the water wave, something strange happens. The water waves will no longer pass through this small opening and carry on as if nothing had happened. Instead, the small opening acts surprisingly like a new source of waves and sends them out in a wide circular arc on the harbour side of the seaway as shown

19

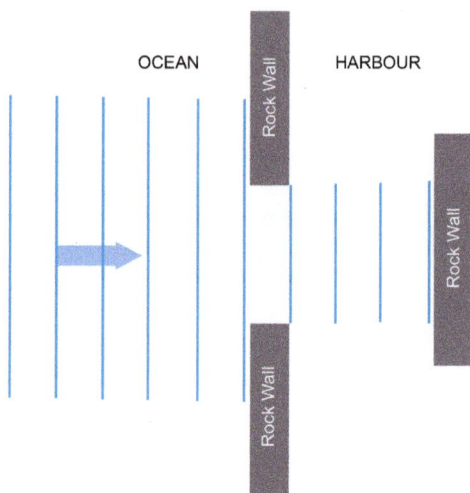

FIGURE 3d: Effect of seaway width on incoming waves. For a wide harbour mouth (relative to the wavelength), the waves pass through the harbour mouth truncated but otherwise unaffected.

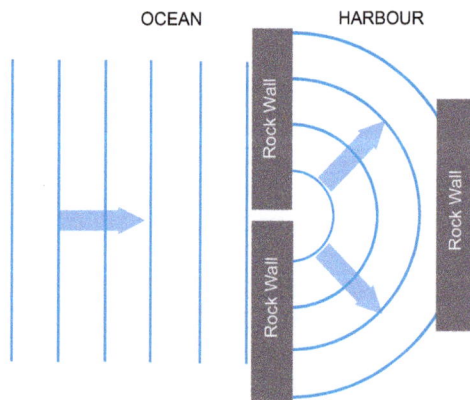

FIGURE 3e: Effect of seaway width on incoming waves. For a narrow harbour mouth (relative to the wavelength), the waves re-radiate from the harbour mouth.

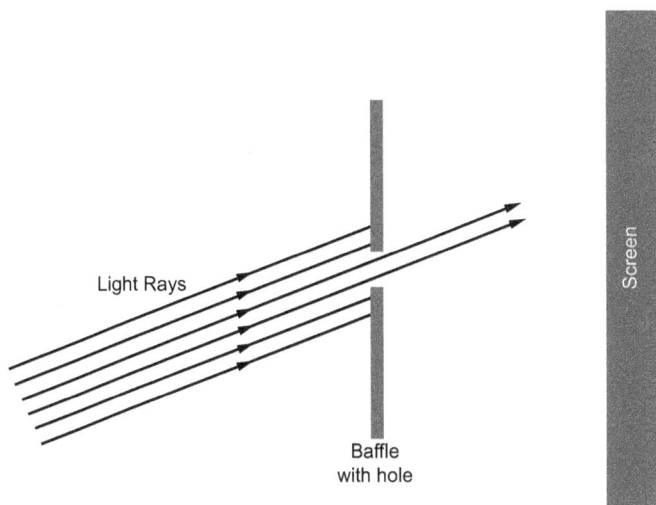

FIGURE 4: Using the particle theory (with light rays rather than wave crests) makes it difficult to demonstrate the effects of diffraction of light through a small hole.

in Figure 3e. This was a very unexpected finding and the phenomenon became known as diffraction. But probably more important than its name is the fact it cannot be demonstrated using the particle theory of light. The particle theory suggests that those light particles that get through the tiny gap will still travel in the same straight line along which they approached the opening. This is shown in Figure 4. If you throw a ball at a wall with a very tiny hole in it, those balls that do go through still travel in straight lines on the far side of the wall just as they would if they went through a larger hole. Under the particle theory, the only difference between a small and large hole is the width of the beam on the far side of the wall and the number of balls which get through the hole.

Light passing through very small holes or slits behaves as waves on water do and re-radiates the light which entered the small hole diffusely in all directions from the far side of the hole, so producing just a broad glow on a screen behind the hole rather than a small intensely bright spot as predicted by the particle model whose particles continue to travel in a straight line on the far side of the hole.

It should be stated here that ocean waves are probably not the best example of these phenomena but they can be seen to illustrate most things so far discussed reasonably well. A better example or model are the small ripples on a very still pond.

Having determined that a very small hole would re-radiate waves just like a new source, Young set up a very narrow slit in a baffle near a light source such as a candle. There was no electricity then. He used a slit rather than a hole in the baffle but the effect is the same and this arrangement could easily be set up in a quiet pond with a dropped stone sending out concentric wavelets just as a candle sends out light waves. This single slit ensures that any light re-radiated from it will be in phase—that all the troughs and peaks of individual waves diffracting from it are lined up or synchronized. All waves start out from the slit in phase. This is important because there is no guarantee that light coming directly from the candle will be like this. A candle's flame is large so one light wave might be emitted from the front of the flame and another from the back or side—all at different distances from the slit. These waves will arrive at the slit in all sorts of phases. The slit knocks the waves into an ordered, coherent, or in-phase beam of light by acting as a re-radiating point source. Today, a laser beam would be used to achieve this coherent wave because that is what a laser beam is.

Young then introduced a second baffle, this time with two slits so that light re-radiating from the first slit fell on the panel with two slits. In turn, because these two slits were also very narrow, they acted like new sources and diffracted or diffusely re-radiated the light which fell on them. The arrangement is shown in Figure 5. For simplicity, let us assume that the candle only emits light of one wavelength, say orange light. Generally, a candle will emit a mixture of wavelengths with some red, yellow and even green and blue light. We can make sure only orange light is used by inserting a filter between the candle and the first slit. In this way, we only need to deal with light of one wavelength and the resultant effect is much more striking if we do this.

Now we have a new situation whereby two nearby sources are radiating light waves which at some point must overlap or interfere as

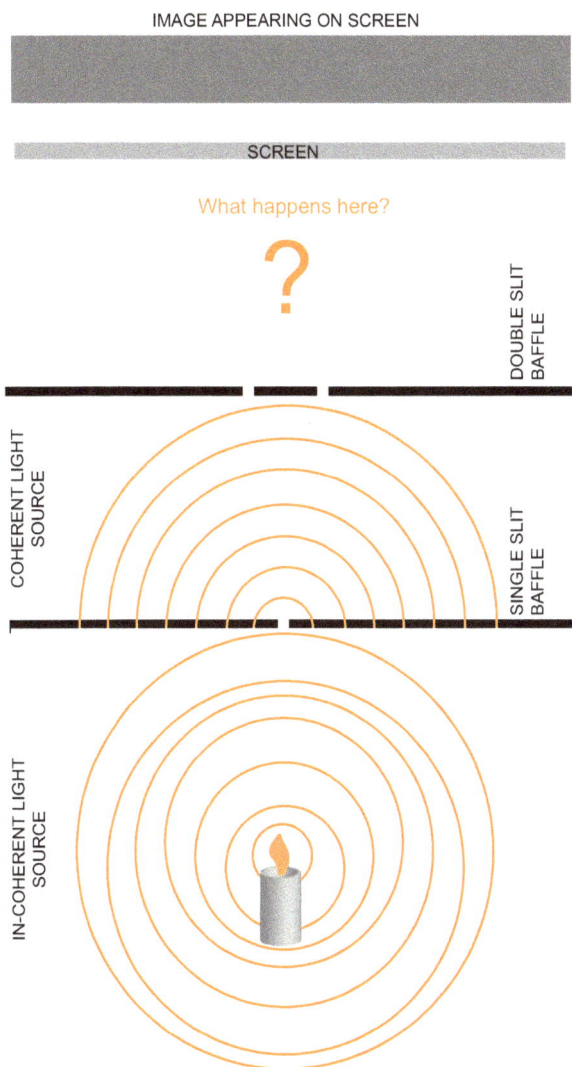

FIGURE 5: Setup for the classic double-slit experiment. What image will appear on the screen after the coherent light passes through the double slits and falls on the screen?

physicists call it. What will happen? On first impressions the answer seems quite obvious. Two sources working together must produce overall a brighter beam and hence a brighter diffuse region on the screen behind the two slits. But this is not what is seen.

Think about water waves. If you have ever been in a wave pool where water waves are generated artificially you may notice some strange things happening. Firstly, you probably see waves reflecting off the sides of the pool as expected but when these rebounding waves strike another wave approaching the wall, there is suddenly a tremendous build-up of wave height as if one wave is added to the other. This is exactly what is happening. If two peaks of a wave meet, they reinforce each other and produce a wave twice as high. If two troughs meet, they produce a very low or shallow region. Alternatively, if a wave crest meets a trough, they work against each other. The crest tries to raise the water level while the trough tries to drag it down. The two effects cancel each other out and the water surface doesn't move at all—it stays momentarily flat. I say momentarily because waves are continually moving and so these effects do not stay in one place for long.

Waves of all types behave in this way as shown in Figure 6a where crests reinforce crests, troughs reinforce troughs and crests cancel out troughs. The resultant wave is produced by adding the height of one wave with that of the other so that, for example, when two crests occur at the same time, the resulting combined wave is twice as high. Likewise, a trough with negative height will cancel out a crest with positive height if they meet, resulting in no wave at all as in Figure 6b.

Armed with this new knowledge of wave theory, it is now possible to deduce what will happen in the case of the two-slit experiment by studying the pattern in Figure 7. Rather than just a bright fuzzy glow, the screen shows distinct bands of light and dark regions. In the centre of the screen is a bright band denoted by A. Here the waves from the two slits arrive along paths of equal length so the waves are in phase. This means that the crests arrive together and so reinforce each other. Behind the crests are the troughs which also arrive together and so reinforce. What arrives at A is a very strong wave made up by adding the amplitudes of the two waves in the manner shown in Figure 6a.

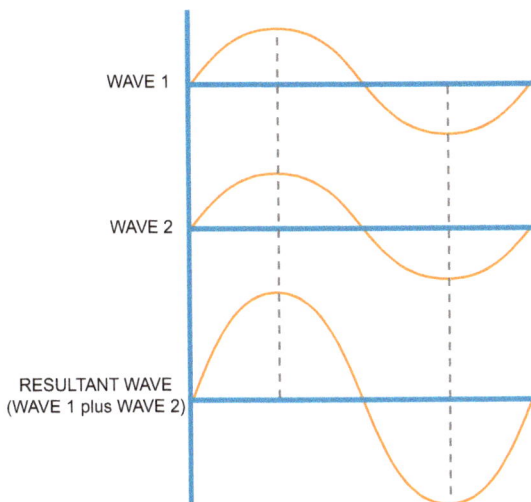

FIGURE 6a: The resultant of adding two waves in phase. Two crests or two troughs added together reinforce, producing a larger amplitude wave. At any point along the x-axis, summing the amplitudes of the two waves gives a value for the resultant amplitude.

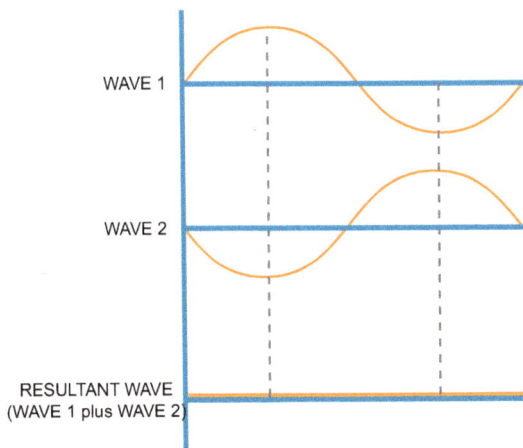

FIGURE 6b: The resultant of adding two waves which are out of phase. A crest and a trough which are aligned cancel out, producing no wave at all (zero amplitude). At any point along the x-axis, summing the amplitudes of the two waves gives a value for the resultant amplitude, in this case a constant value of zero.

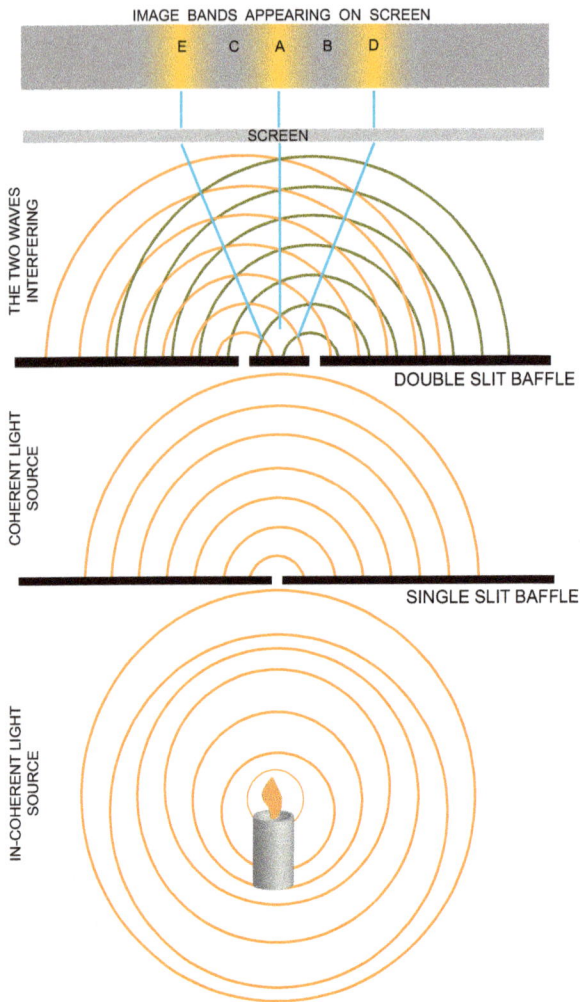

FIGURE 7: Setup for the classic double-slit experiment showing the formation of the diffraction pattern of consecutive bright and dark fringes on the screen due to the interference of the two waves emanating from the double slits. The circular lines indicate the top of the wave crests.

Now let's study what happens either side of the central point A of the screen. At B, the path from one slit is shorter than the path from the other slit. Likewise, at point C. Having paths of different length means the waves will arrive out of phase and, at B and C, the phase is such that a crest from one slit coincides with the arrival of a trough from the other slit. The two waves therefore cancel each other out, as shown in Figure 6b. They interfere destructively, to use the correct terminology, and we obtain a dark band on the screen.

Now if we move still further out from the centre of the screen, we come to points D and E. Again, the paths from each of the slits are different in length but the difference in length is exactly equal to one wavelength—the distance between wave crests. So again, two crests arrive simultaneously even though one of them is a full wavelength behind the other due to the difference in path lengths. This does not matter. What matters is that two crests arrive together and these reinforce or interfere constructively to again produce a bright band. And so the pattern repeats itself as you move further away from the centre of the image.

The double-slit experiment once and for all showed how light behaved like a wave and in particular showed how light waves can interfere with each other—a phenomenon which the particle model could not describe. We see this diffraction phenomenon around us every day in the thin films of oil we find on the surface of pools of water when we try to hose an oil slick away. These colours are produced by a compound version of the double-slit experiment where the two interfering rays come from the top and bottom layers of the film of oil. Rather than seeing light and dark fringes we see the whole gamut of colours from the rainbow because rather than simply using light of one colour as was used in the double slit experiment, in this case we are using sunlight composed of many colours, each with a slightly different wavelength.

Using only a single wavelength as we did in the double-slit experiment means that all the light at one point will drop out due to destructive interference and we will obtain a dark band. With many wavelengths present only light of a particular wavelength will cancel out at a point so we will see white light minus that wavelength at that point. So the light there will no longer be white but biased towards a colour. At another point a different colour will drop out and so on, thus creating the rainbow effect.

Double Slits and Quantum Physics

Although the double-slit experiment was performed in the early 1800s it took some time to displace Newton's corpuscular or particle model. However, by the turn of that century the wave theory was well established and its future seemed rosy. The particle theory was relegated to the scrap heap of history. But then, suddenly, two prominent physicists made two very bold suggestions which refuted the wave theory. These new proposals turned out to be the very stepping-stones to the new era of quantum physics although no one realized that at the time because the discipline of quantum physics was yet to be invented.

New theories only need to be developed when the currently accepted model will no longer predict some new discovery or behaviour. Generally, an attempt is made to extend the existing model to incorporate the new findings but, when this cannot be done, the old theory is thrown out and a new one takes its place which does describe all the previous knowledge plus the new bits which caused the controversy in the first place.

This is what happened with the development of the theory for light. Initially, Newton's particle theory described everything. Then diffraction was discovered and it was found that a wave theory could explain this together with all the previously discovered properties of light. Newton's old particle theory was thrown out. Now as the twentieth century was dawning, Max Planck (1858 – 1947) and Albert Einstein (1879 – 1955) were boldly suggesting once again that a particle theory of light would be required to describe a new aspect of light.

We now have a very odd situation. It is not normal for a scientific theory to revert to an older view. Old theories are generally discarded and never again resurrected. Here, over several centuries, we have continuing confusion as to whether light behaves as a wave or as a particle. Light is a particle that behaves like a wave. We now accept this wave-particle duality. What we don't know is what is doing the waving. To this day, there has been no resolution of this dilemma.

In fact to this day both theories co-exist simply because no one has yet come up with a single idea or model which can describe all the aspects of light. In the early twentieth century, Planck and Einstein were again suggesting the validity of the particle model but it was a particle model far removed and far more advanced than Newton's corpuscular theory. Both the wave and particle theories co-exist today because they are both needed to describe all the properties of light. The wave theory is better at explaining interference, as we have seen, while the particle theory has become better at explaining some of the strange, or as Einstein put it, *spooky* quantum aspects of light.

It is also important to note that light doesn't sometimes behave like a wave and at other times behave like particles even though it is often written about in this loose fashion. Light is both behaving as a wave and a particle at the same time. The problem is that no one has yet thought of a metaphor to describe such behaviour. So, dear reader, feel free to suggest one.

Let's continue tinkering with the double-slit. What, for instance, happens if we close one of the holes? We find that the interference pattern disappears and we simply get a dull uniform glow on the screen—no zebra patterns here. Open the closed hole and close the other and we get exactly the same dull uniform glow. Open both and again the light and dark stripes appear because once again we have two interfering beams. At first it seems odd that opening two holes does not double the amount of light but we have explained why this is not so in a very acceptable and believable manner using basic wave interference theory.

Now let's switch theories to the modern particle version of light which says that light comes in little packets called photons which cannot be broken down or divided up any further. In this particle language, we can increase the intensity of a light beam by sending more photons. In contrast, in the alternative wave theory, we brighten a light beam by increasing the wave's amplitude or height as shown earlier in Figure 1.

Today, it is possible to create a light source whereby we can very accurately control the release of the light photons so effectively that only one at a time is released. The ability to do this opens up all types of possibilities for exploring the real meaning of wave interference. Think of it. If we replace Young's candle with this modern single-emitting photon source, we can fire one single photon at a time towards the single slit. This photon will emerge on the far side of this slit, diffracted at some angle and perhaps then enter one of the double slits and thereby travel on towards the screen, striking it and briefly illuminating a small spot. Over a period of time, after we have fired many single photons, we could note the distribution of the strikes on the screen to see where most photons are striking.

Now here is a question worth pondering. Can our double slit apparatus still show interference at work if we only emit a single photon at a time? Common sense and logic tell us definitely not. How can there be interference between one photon passing through the two slits of the apparatus at a time? Interference requires two players to tango simultaneously.

To answer the question all we need to do is look at the screen and see where the photons are striking. If there is interference then there will be areas which coincide with the dark bands of the earlier interference pattern where the individual photons never strike. As well, areas where the photons do strike will coincide with the bright bands of the earlier interference pattern. If interference is nullified by only releasing one photon at a time then we would expect the photons to strike anywhere on the screen—in a pattern similar to that obtained earlier by closing one of the slits, so producing a diffused general glow across the entire screen.

Common sense tells us that firing only one photon at a time nullifies the possibility of interference occurring because two waves are required for interference but common sense is not always correct and it certainly isn't here. Strangely, we do observe an interference pattern. It is almost as if the photon, which is a single entity that cannot be split, actually does split, with half of it going through each of the slits and interference occurring between the two halves of the photon at the screen.

In the centre of the screen we observe many strikes as each photon is fired towards the double slits. Here, as before when using a much stronger beam of many photons, we have constructive interference between the two beams. This constructive interference repeats itself at regular intervals out from the centre of the screen when the two beams are in phase. Between these bands, destructive interference occurs where the two beams cancel each other out because they are out of phase, and the screen is dark. As expected, when we fire only one photon at a time these regions never receive a hit. It is as though the single photon again interferes with itself and, this time, cancels itself out of existence. The single photon is observed leaving the light source but it never turns up at the screen.

There is more. The strange story does not end here. It is possible to attach sensors to each of the double slits which can indicate if a photon passes through. This then will decide whether the photon splits. If it does, both detectors will go off. But oddly it is found that both detectors never go off together. Only one of them ever fires so the photon really does go through only one of the double slits and which one it chooses is quite random. There is no way of predicting.

The single photons do not split. As well, with the detectors in place, we now find the interference pattern has disappeared! The very act of knowing through which slit the photon passes destroys the interference effect. More importantly, it has been shown that it is not because the detectors interfere with the experiment in some way that the interference pattern is destroyed. It is simply knowing which slit the photon passes through that destroys the interference pattern. This is the confusing part of the quantum theory that we do not understand. The effect of measurement destroying interference is at the heart of quantum weirdness. This is where the quantum theory breaks with any classical physical theory.

This is becoming absurd but such are the strange things of the quantum world, or more precisely, our own world. So strange in fact that many a physicist has metaphorically collapsed with disgust when it dawns on them how nature behaves. Richard Feynman (1918 – 1988) was led to say *nature has got it cooked up so we'll never be able to figure out how she does it.*

These weird quantum effects are not restricted to photons. Every particle behaves in a similar fashion. We could fire electrons or atoms through the slits and get the same results. It is just a lot easier to do the experiment using light. Quantum theory also seems to suggest that, under the right circumstances, larger objects such as tennis balls will behave similarly.

But we are getting way ahead in the development of the quantum theory. These weird realizations came much later. In the early days of quantum physics, at the beginning of the twentieth century, the discipline did not even have a name. The field occupied only a very small area in the great discipline of theoretical physics and only a few years prior to that the quantum concept had not even been thought of.

So having explained the wave-like nature of light by way of the double-slit experiment in some detail you might be dismayed to find that it does not have much to do with quantum theory. More correctly, I should say that the wave theory of light does not have too much to do with quantum theory because just when the wave theory had become firmly established in the scientific community, along came Einstein again to blow away these misconceived ideas.

In 1905, Einstein proposed a reversion to the particle theory of light to explain the photoelectric effect but in doing so he was not suggesting a complete re-adoption of Newton's corpuscular theory. Einstein's idea would turn out to be a lot more robust, but what he proposed in 1905 was the mere beginning of a new particle theory of light which would eventually develop into the full quantum theory and explain things the wave theory of light could never hope to. Let's begin with Einstein's 1905 suggestion.

CHAPTER 3

THE NEED FOR CHANGE

The quantum theory developed from a discovery that the values of some things in classical physics which were considered continuous, or had a continuum or smooth range of values, were in fact quantized. This simply means that the values of certain commodities had to be specific values and could not be just any value. The weights of potatoes for instance are not quantized. Within reason, a potato can be any size and weight. In contrast, a cricket ball or baseball has a specific size and weight. It can only have one allowed weight and size. It can be said that the weight and size of these balls is quantized, albeit into one category only. Other items may have several categories—for instance, shoes come in a dozen or more different sizes. If one size is just too small then usually the next size will fit assuming the jump between sizes is not too big. Shoe companies limit the number of sizes for economical reasons—it would not be practical to offer a hundred different sizes and in fact it is not even necessary. Shoe fit does not need to be that exact.

In early science prior to the twentieth century, the idea of quantization was quite rare and, in the classical era, quantized items were few and far between. The mass, weight, velocity and energy of a body could be any value at all. No value was forbidden. In the classical era, quantized values were a foreign idea because nobody had thought there would be a need for them. Nothing in nature seemed to be quantized but that was only because nobody had looked closely enough or had checked out the fine details.

If we disallow fractions, then the series of whole numbers could be argued to be quantized. We know there are numerical values such as 4.23 or 4.56 between the whole numbers 4 and 5 but if the rules say we cannot use them then we are dealing with a quantized number scale. This is what happens when a dice is thrown. Only whole numbers between 1 and 6 are allowed. A dice therefore could be considered a quantized item.

As it stood at the beginning of the twentieth century, the classical view of nature had problems. Two types of existence had been identified. It seemed that physical entities consisted of either solid objects or they were made up of fields or waves such as was the case with light. Now both particles and waves are described by separate sets of parameters and although it is not obvious here, it requires only a few to describe a particle but an infinite number to describe a wave (7). These parameters can also be described as degrees of freedom.

As well as describing light as a wave-like phenomenon, Maxwell had shown that in a classical world—one that could be described adequately using classical physics, the energy available to a system of particles and waves distributed itself equally amongst all the degrees of freedom. This he called the equipartition of energy (8). It follows from this that since waves have an infinite number of degrees of freedom and the particles only a few, the waves get more than their fair share of the available energy, in fact all the energy. As such, systems containing waves and particles could not co-exist. They do, however, and classical physics could not explain why. As we shall see later, the classical view of an atom composed of a nucleus surrounded by orbiting electrons would also not be stable for the same reason. It would instantly collapse. Classical theory needed significant renovations.

The German physicist, Max Planck (1858 – 1947), was born in Kiel in Germany and studied at the University of Berlin before receiving his PhD. from the University of Munich when he was a mere twenty-one-year-old. After several positions, he was appointed professor of physics at the University of Berlin and remained there until he retired in 1928 at the age of seventy. (5)

In the late nineteenth century, there had been many attempts to theoretically explain the so-called blackbody radiation effect—a term which referred to the radiation of heat or light from an object

Portrait of a young Max Planck, probably in his early twenties about the time he received his PhD.

(AIP Emilio Segrè Visual Archives)

Dr. Max Planck

which was a perfect emitter or absorber of heat. The term black was coined because it was known that dark-coloured objects absorbed and emitted heat better than light-coloured objects. On a hot day at the beach, try walking in light-coloured and then dark-coloured sand and you will get the idea.

Theoretically, a perfectly black object will not reflect any light but instead absorb and subsequently re-radiate all light or heat falling upon it. The problem involved formulating an equation which would describe the observed range of frequencies of this emitted light and the problem became known as the ultra-violet catastrophe. At the turn of the century, classical theory predicted all the heat energy from such a body would go into the radiating field at higher and higher frequencies of light—hence the use of the term ultra-violet which is high frequency light. The existing theory predicted no upper limit to this frequency or energy which of course contradicted what was observed. Classical mathematical theory predicted an infinite amount of energy at the highest frequencies, which is absurd. In fact, observations showed most energy was contained in the radiation at the middle range of frequencies as shown in Figure 8. Again the classical theory was breaking down. No one could find the theoretical equation which matched the observed data.

Progress in science is made when dichotomies between theory and observation are resolved, so requiring an update of our understanding of nature. When theory can no longer explain observation then we are forced to update our understanding of reality. We must make changes to our current paradigm and progress on to new ideas, so these unresolved conflicts are actually the engine of scientific progress.

Many attempts were made to solve this blackbody riddle using the classical equations of electromagnetic radiation (to which light and heat belong). They all failed. Then, in 1900, Planck stumbled on a radical scheme. He found that if the energy of the light emitted from the black body was confined only to certain values and not any

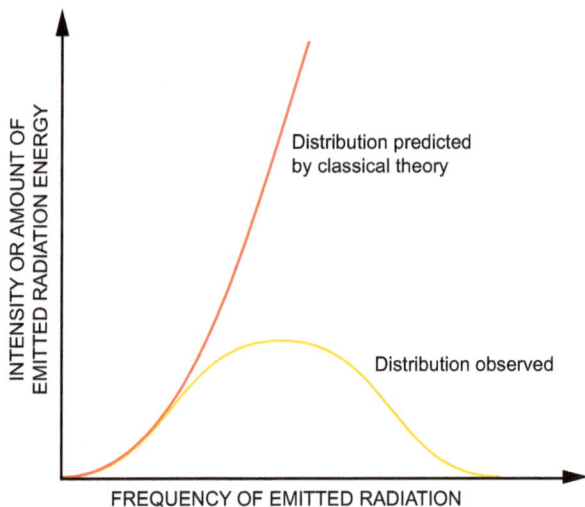

FIGURE 8: *The distribution of heat energy radiated from a "black body" according to the classical theory of heat compared with what is observed.*

36

value, then his equation could explain the blackbody radiation effect. Confining a quantity to certain values rather than any value was a new concept and initially it was looked upon with scepticism but, by doing so, the mathematics fell into place. Many considered it a fiddle to match theory with reality.

It is untrue to say that in this year Planck founded the era of the quantum theory and from that moment the gates were opened to a flood of new ideas. Planck made one major contribution which allowed this new field to develop further. It had long been known that atoms under certain circumstances could emit light. Planck went a step further to suggest that atoms could only emit light energy in packets of a certain specific size which he termed quanta. Packets of intermediate size were not allowed. It was the same as saying that potato chips only come in 250 or 500 gram packs. Packs of 300 grams are not allowed and are in fact never observed.

Prior to this suggestion it was thought that light energy emitted from atoms could come in a continuous range of packet sizes—like chips in 251, 252, 253 gram packs, but this theory caused all sorts of mathematical problems. Ultimately, it did not explain what was being observed and therefore the model needed alteration.

The word quantum is often used today in normal speech in such phrases as *a quantum leap*. Commonly it seems to imply a certain grandness—a quantum leap is meant to mean a large step forward. The word's foundations in physics however relate to very small jumps in value, not large ones. Is this then a scientific word which has been misapplied to everyday situations? Generally, that is taken to be the case but that could be ignoring the true depth of the word's meaning. Taking a quantum leap doesn't mean just making a bigger than usual jump. It actually means jumping over no-man's land into a new arena. The actual jump can be quite small. The fact that the landing area is not smoothly connected to the launching pad is the key to adopting a quantum description of the jump.

Adopting Planck's approach very neatly cleared up several unexplained properties of light emission. You might expect then that this new idea would be quickly accepted but the idea was too revolutionary. It seemed like a mathematical trick—a quick numerical fix, a simple patch-up job awaiting a more conventional explanation—but nothing else was forthcoming. Had it not been for

Planck's high standing in the scientific community, the quantum idea would have been ridiculed out of existence. Most regarded it as a mathematical trick which got the right answer by quick and dirty means and Planck must take some of the blame for this himself. His published papers on the topic give the impression that he too was not happy with what he was suggesting. The calculus he used seemed to imply that he knew the answer he wanted to get and so chose the method in order to effect that outcome. Hardly a convincing way to do research in those times but, in the light of many subsequent breakthroughs in quantum theory, it is hardly out of place in this field. It was hard to accept that Planck's numerical methods actually described what was really happening in nature. It was thought that even though his methods got the right answer, they did not reflect what was really going on in nature. In physics, it is not only important to develop an equation which fits the available data. The equation must also relate meaningfully to the physical process it purports to describe.

As time progressed, Planck's method and formula were found to be useful in other areas of research. In 1905, Einstein used it to describe the photoelectric effect and Niels Bohr (1885 – 1962), another quantum pioneer from whom we will hear more later, applied it to his 1913 theory of atomic structure. Finally vindicated, Planck was awarded the Nobel Prize in 1918 for his brilliant breakthrough.

More precisely, what Planck proposed was that the energy of these quanta or quantized packets of light was related to the frequency or colour of the light that the atom emitted. Red light had quanta of one energy while violet light had quanta of a different and, as it turned out, higher energy. It was clear that the energy of a quantum was proportional to the colour or frequency of the emitted light and so Planck was able to relate, quite simply, the energy of a quantum with its frequency by introducing a conversion factor or constant he called h. He proposed the equation $E = hf$ where E and f denote the energy and frequency of a quantum of light and h is the conversion factor or constant which relates the two values. In time, h became known as Planck's constant and still retains this name today.

It is quite simple using Planck's equation to describe what is physically happening in the real world and why the ultra-violet catastrophe is avoided. In physics, it is always important that any new mathematics that is proposed does in fact relate to some new physical aspect of the physicist's model of nature. So, at the high frequency end of the electromagnetic spectrum, the quantum energies are large since $E = hf$. In other words, a very high frequency leads to a very high energy and only a few of an atom's electrons will have that much energy to release to create an equivalent energy quantum. Because there are only a few electrons within an atom with a large amount of energy, even when combined, the right-hand end of the distribution curve in Figure 8 ultimately returns to zero. On the other hand, for very low frequencies and therefore low energies, there is an abundance of electrons with enough energy to emit a low energy quantum but they carry so little energy that even when statistically combined with the many other low energy quanta, their total contribution is still small, so the left-hand end of the distribution curve returns to zero as well. It is only in the mid-range of frequencies where there is a reasonable number of an atom's electrons with a reasonable energy that the bulk of sizeable quanta are emitted and which, when taken together, carry most of the radiation intensity.

Many scientific ideas when ultimately expressed look amazingly simple as Plank's equation seems to be. How could Planck be awarded a Nobel Prize for such a simple equation? It wasn't so much the equation but the intellectual leap required to think of nature in this way. Most researchers, when they look back on a piece of original research they have just completed, wonder why it took them so long to work it all out. But good ideas do not come every day and any researcher will admit they thought of a lot of bad ideas before somehow finally stumbling onto the obvious answer. Obvious now in hindsight but at the time most obscure.

In 1905, five years after Planck published his paper on quanta, Einstein published three earth-shattering scientific papers as well as submitting his thesis for a PhD relating to the theory of molecules. It was a productive year. One paper had to do with proving the existence of atoms, another with the now famous subject of relativity and a third paper led to him being awarded the Nobel Prize. Einstein might best be known for his work on the space and time bending theory of relativity, but it was his paper describing the photoelectric effect in terms of a quantum theory which most impressed the judges of the time. Like Planck, Einstein had to wait, in fact until 1921, before the significance of his work was fully recognized by the Nobel committee. Yet again the quantum theory was slow to capture the minds of the scientific community.

In the early 1900s, Einstein was just twenty when he read Planck's paper on quanta. He was the first researcher to really take the idea seriously and investigate its potential for describing other events. At the age of twenty-five, Einstein simply applied to light in general the same equation $E = hf$ that Planck had applied to light emitted from atoms and in doing so declared that light energy came in packets and not via a continuous wave as had been thought the case for many years. This was the revival of an idea which Newton had first proposed over two hundred years earlier but, unlike Newton's early attempt, this theory stood up to more rigorous examination.

What Einstein proposed was only subtly different from what Planck had suggested in 1900. Planck declared that the light emitted from atoms came in clearly defined amounts which he termed quanta. He did not say that all of light was composed of packets of energy or quanta. This is the step that Einstein took. Planck had discovered the quantization of light but did not realize his idea could lead to a modern-day particle description of light. The equation that Planck had applied to atoms to predict the type of light they would emit, namely $E = hf$, Einstein now applied to light itself declaring that light is not a continuous wave as had been thought for so long but rather consists of packets, particles or quanta. We now call these light quanta photons. The word was introduced into the language of physics in 1926 by Gilbert Lewis working in California (6). Curiously, Einstein used the wave theory of light in his theory of relativity while at the same time using the particle theory of light for his study of the photoelectric effect.

Was Einstein really suggesting a return to the old particle model of light which Newton had initially proposed? Was the work on a wave theory developed over the past two centuries a complete waste of time? Not at all. Einstein was only suggesting that certain properties of light were best described by a particle theory. The wave theory was still quite valid for those properties which it described well and this is still the situation today. Nobody has yet come up with a model using a single idea which can describe the full gamut of properties of light. The problem with light is that several metaphors are required to describe it completely. Today this is called the wave-particle duality. In some circumstances light is seen to behave as a wave. In others it seems to be a stream of particles.

Einstein's theory began with the research of two independent workers, Philipp Lenard (1862 – 1947) and J. J. Thomson (1856 – 1940). Several years earlier, in 1899, Lenard had discovered that electrons were emitted from a metal plate in a vacuum when light was shown it—the so-called photoelectric effect. Two years before this, in 1897, J. J. Thomson working at the Cavendish Laboratories in England, had discovered the electron. These were indeed pioneering times. Both men were to be later awarded the Nobel Prize for these ideas. Finally it seemed the atom, as had always been thought, was not in fact the smallest indivisible particle. As a matter of interest, the origin of the word atom has its roots in the Greek, deriving from the word atomos meaning indivisible—in hindsight now a total misnomer.

The photoelectric effect involves the emission of electrons from certain metals when light is shown upon them. The energy of the light is enough to eject an electron from the atoms of the metal and the presence and speed of these ejected electrons can be detected and measured by scientific instruments. That light is able to do this is one thing but there is something far stranger about the phenomenon which needed explaining and this is what the quantum pioneers were concerned with.

In order to simplify the process, light of only one frequency or colour was used, as was done with the double slit experiment. With the light source set at a certain strength it was noticed that all the ejected electrons were thrown out of the metal surface with the same speed. Then, when the intensity of the light beamed was turned up, more electrons were ejected with exactly the same speed. The expected did not happen. That is, faster electrons were not observed. No matter how strong the light beam was, all the emitted electrons had the same speed. There was just more of them. Further, if the colour or frequency of the light was changed then and only then was the intrinsic speed of the ejected electrons also changed.

It was thought, not unreasonably at the time, that if more light energy impinged on the atoms of the metal then its electrons would be thrown out with higher velocities. Instead, it was found that for a constant light colour no matter how strong the beam of light, all the emitted electrons had the same speed but, change the colour or frequency of the light, and you changed the speed of the emitted electrons. This was what had to be explained and this is what Einstein eventually did by proposing that light came in packets or quanta with a distinct energy only defined by the colour or frequency of the light. It was these light quanta that connected with an electron and gave up its energy to the electron. If this acquired energy was enough, the electron was tossed out of the metal. Any excess energy from the light quanta, left over after the electron was thrown out, gave the electron some kinetic energy or speed.

This then explained why shining more light of the same frequency onto a metal surface did not produce higher energy electrons. More light simply meant more light quanta of the same energy so more electrons could be ejected with the same speed and kinetic energy, not greater energy since the energy of all the light quanta was still the same.

Planck's proposal of the quantum concept did not open the flood gates to a new and wonderful quantum theory. His research merely offered a peep through a doorway which until then had been firmly locked, or to use a more exacting metaphor, he offered a peep through a doorway which no one knew even existed. Planck was forty-two years old in 1900, the year his paper was published. Although ushering in the new era of modern physics, Planck spent much of the rest of his life at the University of Berlin, attempting to incorporate his quantum ideas into the older structure of classical physics which had served him so well in his earlier research. His political views were definitely anti-Nazi and this meant his life was potentially under threat during the 1930s and 1940s. Planck died in 1947 shortly after the end of World War 2 at the impressive age of eighty-nine years.

Planck clung to the old views of classical physics not because he considered his quantum ideas wrong. Rather he considered all physics to have firm foundations in the classical theories which had been carefully developed over the last few hundred years. These theories had stood the test of time and so he thought that new ideas should be rationalised to fit in with them. He had derived his quantum theory from classical principles and, as such, his quantum packets of light or quanta should be capable of explanation in a classical way. In this endeavour he did not succeed. Perhaps it was a matter of age. Now into the middle of his life, Planck found himself clutching to his classical roots.

Not so the young and gung-ho, twenty-something Einstein. After his very productive year of 1905, Einstein found no need to revert or reconcile his quantum ideas with classical physics. In fact he vigorously applied the quantum idea to several other related fields of physics with just as much success. Exhausting these avenues, he finally returned once again to classical pursuits, his curiosity with the power of quantum theory well and truly satisfied. In 1916 he published his complete general theory of relativity, probably one of the greatest ideas ever to be developed in classical physics.

It was up to a new generation of researchers to pick up and carry the quantum banner into the new and exciting era of the 1920s. By this time Einstein was in his early forties. The 1920s are well known for their roaring lifestyle but few outside the discipline of physics realize that this time was also the most productive era in the new branch of quantum physics.

Niels Bohr (l) and Max Planck (r) probably in Copenhagen lecturing in 1930. (AIP Emilio Segrè Visual Archives, Margrethe Bohr Collection)

43

CHAPTER 4

EXPLORING THE ATOM

It must be remembered that in the 1920s the idea that the atom was composed of several parts was a fairly new concept. The electron was only discovered by J. J. Thomson in 1897. Much earlier than this it had been realized that the elements were composed ultimately of very small particles or atoms with each element having a different atomic weight. The various elements had different properties—some were solids, others were liquids or gases. They also reacted with each other in many different ways. To explain all these interactions it was concluded there must be something very different about the atoms of each element other than simply their atomic weights. The structure of atoms needed to be looked at in more detail. With the discovery of the electron it was realized that the atom was not the fundamental particle of nature. The atom was in fact composed of constituents of which the electron was but one and it carried a negative charge of electricity.

Identification of the components of the atom had in fact begun unwittingly even earlier. Research with radioactive materials, in which both Marie and Pierre Curie were involved, had uncovered some strange forms of radiation being emitted from these radioactive materials. Ernest Rutherford (1871 – 1937), a British physicist born in New Zealand, who had worked at the Cavendish Laboratories in England under Thomson in the 1890s and later in 1898, was appointed professor in physics at McGill University in Montreal. He gave these rays the names alpha and beta radiation. Later, a third type of emitted ray was logically called gamma rays.

Portrait of Ernest Rutherford

(AIP Emilio Segrè Visual Archives, gift of Otto Hahn and Lawrence Badash)

Thomson realized that beta radiation was in fact a shower of negatively charged particles which turned out to be electrons. Alpha radiation turned out to also be a shower of particles rather than light but these particles were far heavier than electrons and also positively charged because they were deflected in the opposite direction to electrons when passed through an electric field built to analyse the weight and charge of these particles. Later it was found that alpha particles were in fact the positive nucleus of a helium atom but that is not of great importance here. Interestingly, of the three supposed rays, only gamma rays turned out to actually be radiation. It was later found that gamma rays were indeed electromagnetic radiation or light of a very high frequency, even higher than that of x-rays. As well, gamma rays do not need to concern us here.

45

What came out of these discoveries was the realisation that atoms may well be composed of both positive and negative particles. If electrons were negative then surely there were positive particles which balanced that charge so making matter in general electrically neutral. The alpha, beta and gamma nomenclature is still topical because even today the terms alpha and beta rays are sometimes used even though everyone knows they refer to particles. And so a quest for a model of the atom's structure began.

The Watermelon or Thomson Model

Atoms are neutral particles, having equal amounts of negative and positive charge. It was discovered when an electron was removed from an atom that the remaining atom now had a slight positive charge because of the imbalance caused by the removed electron. The simple removal of an electron from an atom did not however shed any light on the arrangement of these charged particles within the atom.

Thomson proposed a model similar in structure to a watermelon where the positive charge is spread out uniformly throughout the entire atom like the red flesh of the melon. The much smaller electrons would possibly be scattered throughout this positive cloud like the black seeds. This Thomson model, as it was called, turned out to be wrong but it was a fairly good starting point for further investigation.

The Cherry or Rutherford Model

The cherry model is more correctly called the Rutherford Model but I have used the idea of a cherry with a single central seed to illustrate its make-up and contrast it with the watermelon model just discussed. The key to the discovery of this improved model came in 1909 from the physics department at the University of Manchester to where Rutherford had moved just two years earlier. He had received the Nobel Prize for his work in radioactivity in 1908.

Researchers under Rutherford, including the German physicist Hans Geiger (1882 – 1945), developed a method of firing a stream of alpha particles, derived from radioactive material, onto a thin sheet of metal. Special detectors were developed to sense where these alpha particles ended up. Some went straight through the metal sheet. Others went through but were deflected slightly. Still others bounced straight back like a ball hitting a wall. It was Rutherford's job to try to explain a structure for the metal sheet which would deflect or scatter alpha particles in this way.

Knowing that the alpha particle is positively charged and that like charges repel, Rutherford determined there must be other positively charged particles in the metal sheet. If an alpha particle hits one of these head-on it will bounce straight back. If it passes between two positively charged particles it might go through the metal without being deflected or it might deflect slightly if it passes close to a positively charged particle. These three possibilities are shown in Figure 9.

Rutherford knew there were also negatively charged electrons in the metal but he figured these must be very light compared to the positively charged particles, assuming these positively charged particles were anything like the positively charged alpha particles that were fired at the metal. These light electrons would be no match for the much heavier alpha particles which would simply brush the electrons aside as they passed through the metal.

The watermelon model of Thomson could not explain the behaviour of the alpha particles. Using this model containing a uniform cloud of positive charge inside the metal atoms would mean that all incoming alpha particles would be treated equally by the atoms of the metal. There would not be a range of deflections.

Rutherford proposed his new model for the atom in 1911. He suggested there must be a central core which he called the nucleus and which is much heavier than electrons and positively charged. This positive charge is balanced by a number of negatively charged electrons which encircle the nucleus like the planets encircle the sun. Together, the positive and negative charges balance out so producing a neutral atom. Subsequently it was determined that most of the atom is made up of empty space, there being a huge void surrounding the nucleus through which the electrons moved. Even solid objects like metal and rocks are mostly made up of space.

Rutherford's model of the atom revolutionised science but inevitably it would need adjustment because there was one aspect

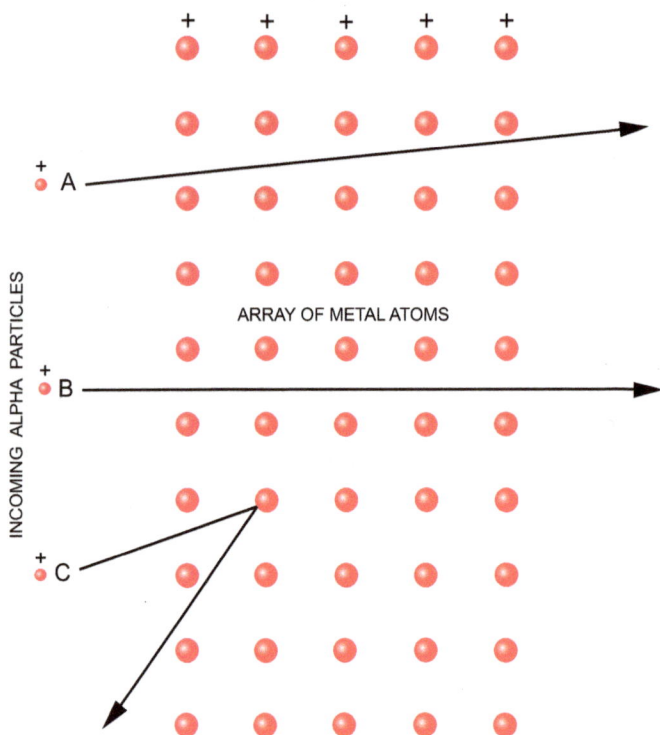

FIGURE 9: Looking edge-on at the atoms of a metal sheet, five atoms thick, being bombarded by positively charged alpha particles (small pink balls). Only the positively charged nuclei of the metal atoms are shown (large pink balls). The negatively charged electrons surrounding the nuclei are not shown as they have little effect on the incoming alpha particles. The alpha particles at A and B pass through the metal relatively unaffected. The alpha particle at C is scattered back from one of the metal nuclei.

of its structure that could not be easily explained. If unlike charges attract, how was it the electrons orbiting the nucleus did not collapse into it. Did the atoms possess some strange force like the centripetal force which balances gravitation so allowing the planets to continue in their orbits around the sun? Were the electrons really orbiting the nucleus like little planets? This part of the model needed clarifying.

Although difficult and lengthy to explain here, Maxwell's electromagnetic theory had predicted that a charged object like an electron orbiting a nucleus must emit electromagnetic radiation or light. Losing energy in this way meant the electron must slow down and if it slows down it will gradually fall towards the central nucleus and finally collide with it. This fact put an end to the similarities between the solar system and the atom. Orbiting planets were not electrically charged and so did not emit energy in the form of light. If they did, they too would slowly lose speed and spiral into the sun. This fact alone meant the Rutherford model for the atom was on shaky ground.

A Modified Solar System Model

First we saw how the positively charged and negatively charged elements of the atom were recognized in the Thomson watermelon model. Then came the Rutherford cherry model—a more exact description of nature which described a positive nucleus encircled by negative electrons. Now came a refining of the behaviour of the negatively charged elements or electrons of the atom which led to the atomic structure according to Niels Bohr. And even though this model as well finally turned out to be mostly incorrect, it is this model which heralded the coming of the new quantum theory of atomic structure and is the one which is still often taught in schools, at least initially, today. The basic atomic model as visualized by Rutherford and Bohr is shown in Figure 10a.

The idea for its design stems directly from Bohr's approach to research. Bohr had the ability to propose somewhat outrageous ideas if they fitted in with what was observed. He didn't worry too much about the physics—at least at first. So, to overcome the problem of electrons losing energy and subsequently spiralling into the nucleus, he simply created a quantum law which said they couldn't do that! To hell with the classical laws of physics. The quantum world behaves differently. Bohr proposed there must be something in the make-up of an atom which forces its electrons to keep all their energy and not gradually give it up.

The clue came from the previous work of Planck who had suggested that the light emitted from atoms came in discrete amounts called quanta and from Einstein who proposed that light itself was composed of these quanta which had come to be called photons. The fact that some quantities, such as the photon energy emitted from atoms, came in discrete or stepped amounts was the key to Bohr's suggestion.

Niels Bohr (1885 – 1962) was a Danish physicist who was awarded the Nobel Prize in 1922. In 1911 he had finished his PhD and gone briefly to work with Thomson at the Cavendish Laboratories, then on to Manchester to consult with Rutherford before returning to Holland once again after several months. In 1913 he proposed his radical model of the atom.

Bohr's suggestion was that the electrons did indeed orbit the nucleus like planets but did not gradually lose energy and spiral in because the electrons were forbidden to lose energy gradually. Under his new quantum theory of atomic structure, electrons were only allowed to emit energy in certain quantities. In other words, in whole quanta. It was the equivalent of saying that a planet such as Mars could not lose energy gradually and so begin to spiral in towards earth's orbit. If Mars were to lose energy for some reason it must do it in big quantum chunks and so drop instantly from its current orbit to that of, for example, earth's orbit, essentially in one step.

What Bohr suggested ran counter to the doctrine of classical physics where quantities such as energy can adopt any value and can change smoothly or continuously from one value to another. Now Bohr was saying the allowed values for the energy of an electron could only be certain specific quantities. Bohr's theory was a mess. It combined certain classical ideas such as orbiting electrons behaving like planets while at the same time incorporating some of the more modern quantum ideas of fixed values or quanta.

No matter how bad was the structure of Bohr's model, it did work. It explained what Planck had observed many years before—that atoms released light in discrete amounts at a certain frequency and only in

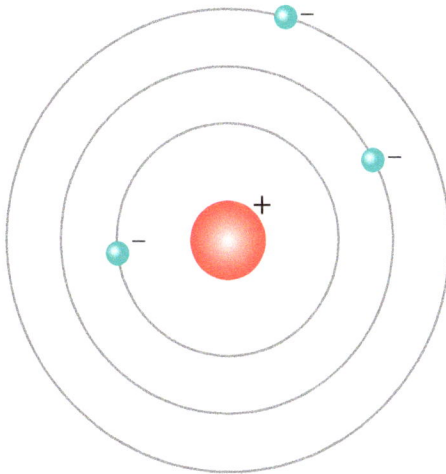

FIGURE 10a: The Rutherford/Bohr model of the atom has a central positively charged nucleus surrounded by negatively charged orbiting electrons.

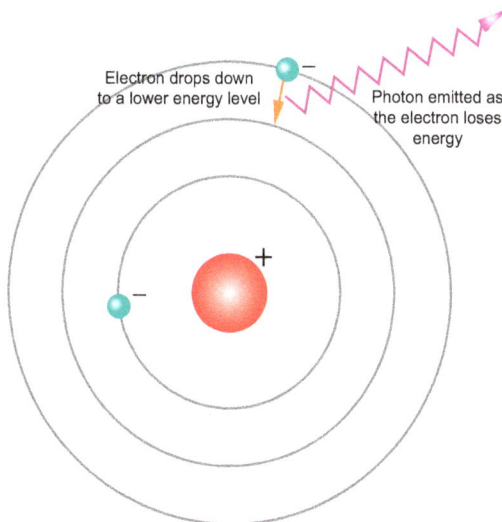

FIGURE 10b: The Rutherford/Bohr model of the atom including Bohr's proposal for an electron to lose energy by dropping down to a lower energy level closer to the nucleus, thus emitting a photon with exactly that lost energy.

certain amounts. Bohr's model explained how this could happen. If an electron jumped down from one energy state to a lower one, the energy of the light emitted would exactly equal the energy difference between these two levels. See Figure 10b. The levels became known as electron energy levels and the gaps in energy between the levels was observed to be characteristic of an element. Different substances emitted light photons of different colour because the electron energy levels were different distances apart.

This much the Bohr model explained but, as was discovered later, electrons do not orbit the nucleus like planets as Bohr had thought but instead occupy a certain confined region outside the nucleus and in this region follow no clear path. They in fact follow a strange unpredictable pathway as they dart around within their allotted region or shell surrounding the nucleus. Most atoms have many electrons surrounding their nucleus and these electrons are confined to their respective shells. These shells can best be thought of as the layers of an onion, one beneath the other, with the electron in each tending to form a blur or cloud because it is moving fast and unpredictably within its allotted shell. These clouds of electrons around the central nucleus is what gives an atom its shape or form.

We often try to think of the structure of the atom in the terms of classical physics—things we are familiar with like a miniature version of the solar system. These models explain some of the properties of real atoms but certainly not all as we have seen. It is now generally accepted that thinking along these lines in an attempt to understand quantum physics is not necessarily a good idea and in fact it may slow down the development of the subject. Many quantum physicists today think in very abstract terms and make no attempt to relate their ideas to familiar concepts in the classical world. Others try to hold on to classical models, adapting them where possible to incorporate quantum properties. Retaining some degree of classicality can help contrast the difference between the classical and quantum worlds.

Having said that, there are some aspects of the atomic structure which do lend themselves to explanation with familiar models. The positioning of electrons around the nucleus is one example. Electrons occupy what are called energy levels around the nucleus. If we ignore what we think might be the physical shape of an atom, namely an onion with layers and a central nucleus, and simply focus on the idea of electrons sitting in energy levels then it is possible to design a semi-classical model of the atom which describes this hierarchy of possible energies. The energy levels can be thought of as little ledges around the sides of a deep well with the nucleus at the bottom of the well as depicted in Figure 11. Bohr introduced the idea of quantum numbers, a series of integers 1, 2, 3, 4 . . . , one of which was associated with and helped describe each energy level. So, the lowest energy level had a quantum number of 1, the next 2 and so on.

Amongst other things, the photoelectric effect studied by Einstein can be explained with this diagram. A photon striking an atom in a sheet of metal gives up its energy to an electron on one of the higher ledges and, with this extra energy, the electron is able to throw itself out of the well and hence out of the atom and so out of the surface of the metal.

Sometimes, however, the electron may not receive enough energy from the photon to actually be ejected from the atom. Instead the electron only jumps up to a higher unoccupied energy level. The atom is then said to be in an excited state because the electron is now overly energized and at a higher energy level than it normally is. After a short time, this electron will drop back down to its normal lower position within the atom. The energy it loses in doing so goes into making another photon whose energy is exactly equal to the energy difference between the two ledges. Think of the electron as a ball which drops further down the well onto various ledges. As it does so, it picks up speed and so kinetic energy. When the ball lands on a lower ledge it stops and its kinetic energy is released as heat and sound as the ball strikes the ledge. The electron can be thought of as doing the same but when it comes to rest on the lower energy level it radiates its kinetic energy as a photon or flash of light. It has been found that by exciting the electrons of atoms of various substances in this way, these substances can be forced to emit light when they normally would not.

Now Planck's formula $E = hf$ relates the frequency f of the emitted light to its energy E and this energy of the emitted photon is in turn equal to the energy given up by the electron in going from a higher ledge to a lower ledge—in other words, the energy difference between the levels due to the difference in height in the model. So, for one type of element whose atoms all have the same gaps between energy levels, we would expect photons of similar energy and hence frequency to be emitted. If all photons have the same frequency, the light emitted is all of one colour. This explains why different elements emit light of different colours and why an element can be identified by the colours in the unique spectrum of light which it emits.

The process is slightly more complicated than this but the essence of the above discussion is correct. Elements can be solids, liquids or gases and this does affect the type of spectrum they emit. As well, a single element may not just emit one colour, hence the use of the word spectrum to imply many colours. Electrons may not simply drop to the next ledge below them but to the second ledge below them, so radiating a photon with a larger energy and hence a higher frequency and different colour. And so on. Most atoms have a multitude of electron energy levels, so producing many different colours in their emitted light spectrum.

The practical applications of these characteristic spectra are enormous, both on a small scale where a substance is analysed in a laboratory to determine its identity, to extremely large-scale applications examining the composition of distant stars. The colour of the emitted light from stars can be analysed to determine what elements produced it even though that star may be many light-years away.

We are all familiar with the rainbow and its colours. A rainbow is formed when white light from the sun is split up into its components by raindrops acting as myriad prisms. Images of rainbows can be easily produced on a screen in the laboratory or even in the home by shining white light through a glass prism. These full rainbows range in colour from deep red to violet and indicate that the light from the sun is composed of many frequencies.

Spectra are used all the time in the laboratory to analyse substances. Unknown substances can be induced to radiate light by first exciting their electrons to higher energy levels and then observing the spectrum of light radiated by the substance when these electrons again drop back down to their normal levels. Rather than a continuous spectrum with bands of colour from red, through orange, yellow, green, blue, indigo and violet, pure elements, in particular, radiate only a few

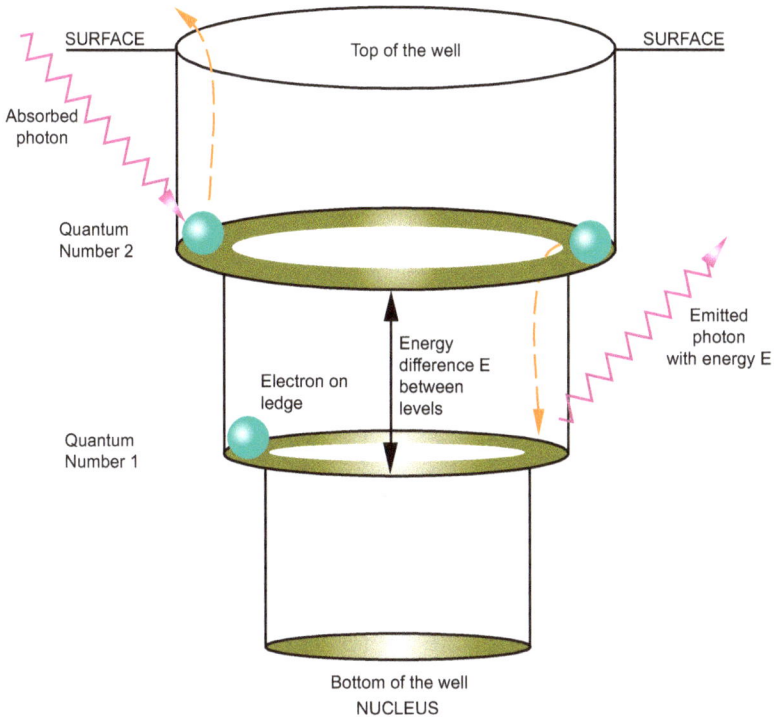

FIGURE 11: A simple model or simulation of the energy levels for electrons within an atom with the lowest energy level or ledge at the bottom of a well and the highest level at the top of the well or at its "surface". When electrons drop down to a lower level of energy within the well they release the difference of energy by radiating a photon of light with equivalent energy. If electrons absorb an incoming photon's energy they jump up to a higher energy level within the well. If they are already in the highest level, the energy of the absorbed photon allows the electrons to escape the atom altogether.

colours depending on the energy differences between their atom's energy levels.

Hydrogen for example has four narrow bands in its spectrum with large gaps between them. There is a narrow band of red, then a big gap in the spectrum with no colour at all until the blue region where there is a narrow line, then a further gap to two lines in the violet region. Hydrogen always radiates this spectrum and can be easily recognized by it. It is its signature and all the other elements have a characteristic pattern to their spectrum as well.

So, irradiating the surface of a substance with light can cause electrons in that substance to be ejected from their atoms altogether or simply to jump up to a higher energy level. Those electrons which go only to a higher level eventually drop back down to their usual level. The exact moment they do this however seems to be totally unpredictable. There seems to be no way of determining when an excited electron will drop back down. It can only be said that it will eventually do just that and this is one of the dilemmas of quantum physics.

Let's try to re-model the atom knowing what we now know. Returning to the earlier attempts to design a model of the atom depicted in Figure 11, the analogy of a well is sometimes drawn upside down so that it resembles a wedding-cake as shown in Figure 12a. This amounts to the same thing but still does not actually model the idea that excited electrons can drop down at any time. One solution is to slope the ledges as in Figure 12b. This certainly ensures that the electrons will drop down but it has problems. Firstly, it ensures that all electrons will eventually drop down (which is not the case in reality) and secondly it predicts exactly when an electron will drop down because we could measure the slope and length of the ledge and determine when an electron reached the edge and fell off. This does not describe the quantum feature of unpredictability—not knowing when the electron will drop off the edge.

Figure 12c solves the problem. By returning to the model in Figure 12a all that needs to be added is a means of randomly shaking the whole structure very gently from side to side. Then electrons will eventually find their way to the edge of a ledge and fall off, but when this will happen cannot be easily determined, so the unpredictability

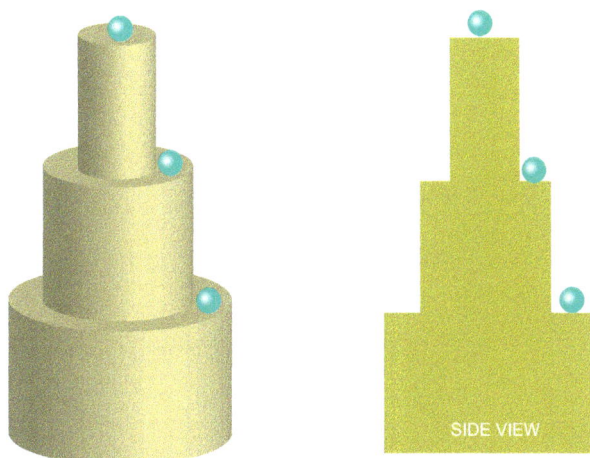

FIGURE 12a: A similar electron level model to that shown in Figure 11 but with the well inverted to form a tiered cake. Either model is a start to illustrating the electrons of an atom but neither allows for electron transitions between levels as the electrons are trapped on the horizontal ledges.

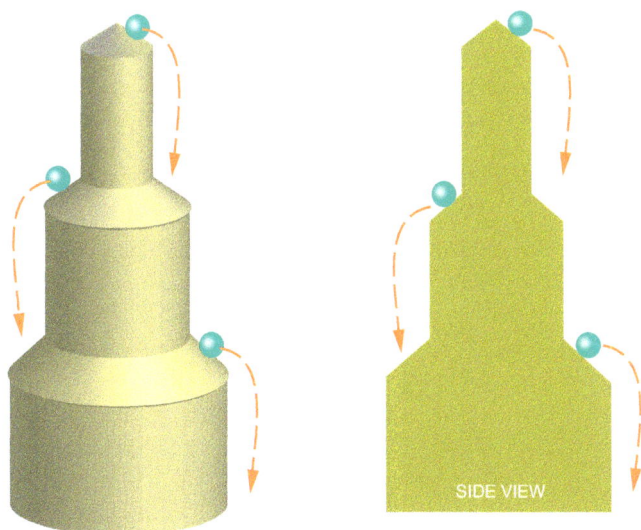

FIGURE 12b: The same electron level model to that shown in Figure 12a but with the ledges sloping to ensure an electron does drop down with assured certainty and at a predictable moment.

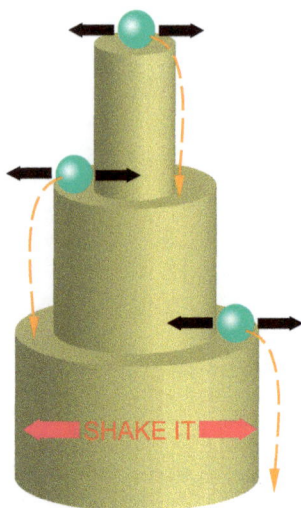

FIGURE 12c: An improvement in the model shown in Figure 12a. Now the model is gently shaken so that at some time the electrons will migrate to the edge and fall off. This is a better simulation of the real world than the deterministic models shown in either Figures 12a or 12b. It attempts to introduce a random element.

of quantum theory is modelled. Scientists spend a lot of time trying to imagine setups like this.

The Bohr model went on eventually to explain much about chemical reactions and the periodic table of the elements. In fact, it did a superb job of explaining the whole structure of chemistry, chemical reactions and the structure of the nucleus, and these findings are still used today. Bohr's contribution was tremendous and he received the Nobel Prize for physics in 1922, the year after Einstein.

It has been said (2), and it is a very good point, that nineteenth century physics is all we need to know to deal with everyday situations. Physics up to about 1923, the time of the Bohr atom, is sufficient to explain and predict chemical reactions and, in fact, most of chemistry. In contrast, the quantum physics developed through the 1930s is enough to keep people pondering to this very day. Although advances have subsequently been made, they only add incrementally to the foundations of the theory developed through the late 1920s and 1930s.

The establishment of the Bohr atom with its quantized energy levels marked the end of the old quantum theory, as it became known, which specifically dealt with atomic structure. The theory was old not so much in that it was subsequently replaced with something better. Rather, the quantum ideas so far developed were extended into a more general theory of physical behaviour, a theory with far more reaching consequences than simply describing atomic structure. Quantum theory was moving into the realm of describing the very way nature was fundamentally designed. This extended quantum theory became known as quantum mechanics because essentially it replaced the old views of classical mechanics developed since the time of Newton. The old quantum theory satisfied most physicists and chemists who spent much productive time applying the new description of atomic structure to solve many issues to do with chemical structure and reactions, and still do so to this day. The quantum theory worked.

There was also a group who wished to delve into developing the quantum theory still further to see where it might lead. And what this group found was a treasure chest of delights and dilemmas they could hardly have imagined.

CHAPTER 5

QUANTUM MECHANICS

To many, the university towns of Cambridge in England, Yale in America and Göttingen in Germany are the towns to nowhere. Nowhere, that is, unless you are associated with the universities because the business of these towns is very much all about teaching and research. These are university towns and, as such, all three are remarkably similar. Similar in layout as well as attitude. Students do not come to these universities to worship knowledge but rather to question it. (3)

It was into such an environment that Max Born (1882 – 1970) found himself when he was appointed chair of physics at Göttingen in 1921. Born, a British physicist and son of a professor of anatomy, was born in Germany. Born was a student in 1900 when Planck published his major paper and received his PhD in 1906 about the same time as Einstein. In fact, they were similar ages. Born waited a long time for his Nobel Prize, finally receiving it in 1954 long after many of his own students had received theirs. Born was noted for his thorough and exacting mathematical approach to problems in stark contrast to Bohr's haphazard approach. Both were brilliant but in different ways and both types of approach were required if the secrets to the quantum world were to be cracked.

Like Einstein, Born, in 1921, was now in his early forties and past what many would consider the prime age at which new discoveries are made. But Born had the ability to attract young high achievers to his institute and at Göttingen he did this with style. The relationship was symbiotic. Born stimulated and motivated his young staff and

Portrait of Max Born

(AIP Emilio Segrè Visual Archives, gift of Jost Lemmerich)

in return was himself inspired to greater achievement. During the 1920s, Göttingen became the centre of the universe for anyone interested in the new physics of the quantum theory. Among Born's researchers was a young student, Werner Heisenberg (1901 – 1976) who in the early 1920s was still in his early twenties and who would, in only a few years, make the important intellectual leap which would thrust the fledgling quantum theory which, at the time, merely dealt with atomic structure, into an all-encompassing theory of nature. Göttingen received a kick-start towards this end when Bohr visited the university in 1922 to give a series of lectures on his new quantum theory of the atom.

The classical theory of mechanics had served the world well for hundreds of years, in fact since Newton formulated his basic laws of motion in the 1600s. It applied to macroscopic objects—everyday

61

objects of reasonable size. Failings were found and adjustments were made, as when Einstein comfortably included Newton's theory into his new theory of relativity. Even so, relativity was still based on and considered part of classical mechanics. Now, however, the world was looking inwards to smaller and smaller objects on a microscopic level and at this atomic scale the old classical theories, as we have seen with the Bohr atomic model, had found their limits.

What was now needed was a thorough mathematical treatment encompassing all the facets of the Bohr atomic model. A totally new theory or even a new way of thinking was needed to explain what was observed on this very small scale and it was Werner Heisenberg who ultimately set the ball rolling. This new theory came to be known as quantum mechanics in order to imply a contrast with the classical mechanics view. This new theory was in fact so successful that quantum mechanics replaced classical mechanics, being valid equally for both the microscopic and macroscopic worlds.

The situation was very much like the discovery of relativity. Einstein's new ideas on motion did not mean Newton's ideas were wrong. Einstein's relativity laws were simply more encompassing, being valid for all objects whether they be at rest or travelling near the speed of light. In other words, Newton's laws were a special case and only valid when objects were travelling at everyday speeds. So it was with classical and quantum mechanics. While quantum mechanics was valid for all situations no matter how small or large, the old classical theory of mechanics was only exact for everyday objects of reasonable size.

It is only natural that early theories were made to describe familiar situations. Then, as it became possible through technological invention to observe more remote cases and microscopic cases, the current theory needed to be adjusted. It was the case with relativity and it was the case with quantum mechanics. Relativity expanded the classical theory to include the realm of very high speeds. Quantum mechanics investigated the very small world of atoms and, in so doing, ultimately called for a total rethink of our view of all of nature.

Physics is an exacting science and prides itself with being the most fundamental of all the disciplines. As such, its discoveries and conclusions must be seen to be precise. What Heisenberg proposed was anything but precise, but he claimed that was how nature was. It was difficult for the traditionalists to agree. Heisenberg's idea became known as the uncertainty principle and implied there was a fundamental limit to which intrinsic features in quantum systems could be known simultaneously—an idea we will revisit shortly. We could ever only know so much about the behaviour of atomic particles, and indeed any particle or object, and this knowledge could never be increased by using better equipment which might become available in the future.

Heisenberg, the theoretical physicist, proved this mathematically without any need for practical experiment. In fact, Heisenberg had few experimental skills. To reach this conclusion by theoretical means is truly amazing and it was this finding which gives Heisenberg the honour of often being called the founder of quantum mechanics. As with any endeavour, there are usually several players who contribute greatly to its development and it must be realized that this was the case with quantum mechanics. It was Heisenberg who made the crucial jumps in thought when they were most needed but, before he did this, several crucial and logical steps needed to be made.

The Dual Nature of Light

That many properties of light could be explained using a wave model was well established. Such things as refraction and interference were easily dealt with in this way. Einstein had re-introduced a particle aspect of light to explain the photoelectric effect but this idea of light being a stream of particles or photons took some time to be accepted. Einstein adopted whatever model he thought suited the current problem

Summarising the properties of the photon, the photon is always massless and furthermore cannot be stopped. The speed of light may change in different media (slower through glass than through air) but the photon always travels at the speed of light for a particular medium and this speed can never reach zero. Appendix 2 looks at the mathematics of the particle properties of the photon.

The Dual Nature of Matter

Light exhibits both wave and particle-like properties. Understandably, this duality presented an intriguing question—did any particles of matter also exhibit wave-like properties? A French physics student studying for his doctorate in 1923 took this dual nature of light, being either a wave or a particle, a large step further. Prince Louis de Broglie (1892 – 1987) suggested that if a wave-like entity such as light could sometimes behave like a stream of particles with momentum then perhaps, conversely, what we call real particles or objects could sometimes behave like waves.

De Broglie had difficulty getting his thesis accepted. It was just too radical. Ultimately Einstein was brought in to comment and straight

Portrait of Louis de Broglie

(Academie des Sciences, Paris, courtesy of AIP Emilio Segrè Visual Archives)

away he realized the importance of de Broglie's findings. De Broglie went on to receive the Nobel Prize for physics in 1929.

De Broglie suggested that all the equations which had been developed so far to describe the energy, frequency and momentum of a photon of light could equally be applied to any particle. This of course would mean that any object could be ascribed a wave frequency which at first seems absurd. How could an object such as a tennis ball have a frequency? Those against de Broglie argued that light sometimes displayed its wave nature while at other times it behaved like particles so it was quite right to have two theories for light but balls and other real objects always behaved like balls or particles. No one had ever seen their wave-like character, so what was the point of de Broglie's suggestion?

De Broglie borrowed two equations relating to light and stated that they also referred to every particle in the universe. The Planck formula $E = hf$, relating the frequency of a photon to its energy, was used to proscribe a wave frequency to any object of energy E. Likewise the momentum of an object was also related to its frequency through the Einstein formula $p = E / c$. Replacing the energy E in this equation with hf from Planck's equation meant that the momentum and frequency of a particle could be related through the formula $p = hf / c$.

What made de Broglie think along these lines? It had to do with what Bohr had proposed as his model for the structure of the atom. Remember Bohr described the atom as a nucleus surrounded by orbiting electrons, like little planets, with each electron orbit confined to a certain allowed energy. Now Bohr went a step further and described the values of the energies that each of these levels were allowed to adopt as well as calculating the electron's angular momentum for each of these orbits. Any object moving in a circle or orbit has angular momentum about the centre of the circle.

Now, the momentum of a body as we saw before depended on its velocity and mass. Travelling at the same speed, a cricket ball has more momentum than a tennis ball and we can feel this when we try to catch each of them. On an ice-skating rink, we would have trouble stopping an overweight skater who had lost control and was coming straight towards us but we could quite easily stop a thin or small skater. These are examples of momentum in everyday life.

More correctly, they are examples of linear momentum because the objects are moving in straight lines. Momentum is in fact just a more technical name for inertia—the tendency of a body to continue in its state of motion or to remain at rest.

During interactions, momentum like energy, is always conserved. When a billiard ball strikes another, the momentum of the first ball is now split between the two moving balls. Each will move more slowly than the original first ball because the momentum must now be shared between two masses rather than one.

An object can also have angular momentum. This is momentum or inertia due to its spinning. A spinning top has angular momentum as does a spinning skater. It depends again on the mass of the object and its angular or rotating speed rather than its linear speed in a straight line. Angular speed simply means the number of revolutions per second. Just as important is the way the mass is distributed within the rotating object. For a given amount of momentum, more compact masses spin faster. Have you ever noticed how a spinning skater speeds up if she draws in her arms then slows down again as the arms are extended? This is simply the conservation of momentum law at work. Her angular momentum does not change but since the compactness of her mass does, to maintain the same momentum the skater speeds up so making more revolutions per minute.

More to the point here is angular momentum due to a mass orbiting a central point like a planet around the sun. It depends on the mass of the orbiting object as well as the distance from the centre of the circle. The further away from the centre the slower the mass will orbit. Essentially, it is the same principle as the spinning skater but with all the mass of the orbiting object concentrated at a distance away from the centre of the rotation. It is similar to twirling a weight tied to a string around your head. If you allow the string to wind up around your raised arm, the rate of the revolutions will increase as the string shortens.

Now part of Bohr's theory of the atom required that the electrons orbiting the nucleus at different levels not only have specific energies as determined by their energy level but also specific angular momentums given by a series of values, namely 0, $h/2\pi$, $2h/2\pi$, $3h/2\pi$, $4h/2\pi$ ... (9). Each of these values is simply the same value, $h/2\pi$, multiplied by the consecutive whole quantum number series 0,

1, 2, 3, 4 . . . The constant h again is Planck's constant which seems to appear in all sorts of places in quantum mechanics and π is simply that universal number (equal to approximately 3.142) that appears in all sorts of formulae including those for the circumference and area of a circle. The above quantum series of values is often written as simply nh/2π where n can represent any whole number.

De Broglie got his idea that real particles, that is particles other than photons, may also have wave characteristics from the fact that electrons, which were at the time considered real particles, were associated with energy levels and angular momentum values dictated by a series of whole numbers—the so-called quantum numbers. Now, in classical physics, it is rare to find phenomena described by a series of whole numbers except perhaps the oscillations associated with wave theory, and wave theory is often explained using the example of musical instruments.

Stringed instruments such as guitars and violins make their sound through the vibration of their strings, but this is not any vibration. Strings are forced to vibrate with certain wavelengths depending on the length and tautness of the string. In fact, a string will vibrate in many modes with many frequencies, but by varying the tightness or length of the string it is possible to make the string more prone to vibrate with a certain frequency. The rule is that the length of the string must fit a whole number of wavelengths as shown in Figure 13. This phenomenon can also be simulated using a length of rope tied to a post and shaking the free end at various frequencies.

De Broglie, using the model of the atom which Bohr had developed ten years earlier in 1913, noticed that the series of numbers depicting the number of wavelengths fitting the length of a vibrating string (1, 2, 3, 4 . . .) looked very much like the quantum numbers for the atom and so envisaged the electron in a specific orbit around the nucleus to be a wave-like entity encircling the nucleus as depicted in Figure 14. The circumference of a circle is 2πr where r is the radius. In the example shown in Figure 14, the circumference happens to have a length of eight wavelengths (count them), so this can be equated with the formula for the circumference,

$$2\pi r = 8 \; wavelengths.$$

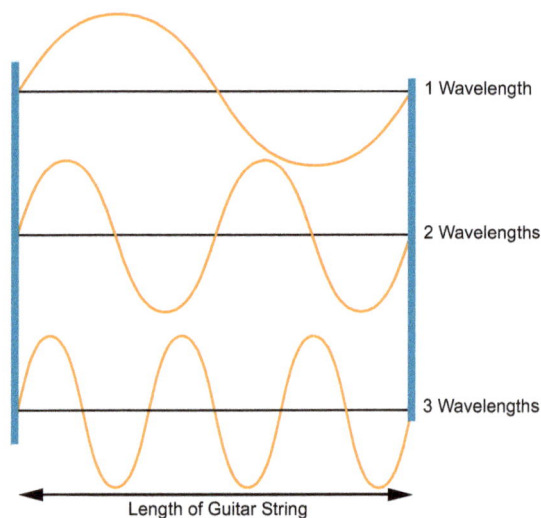

FIGURE 13: Modes of vibration for a guitar string. A whole wavelength or multiple wavelengths must fit within the length of the string.

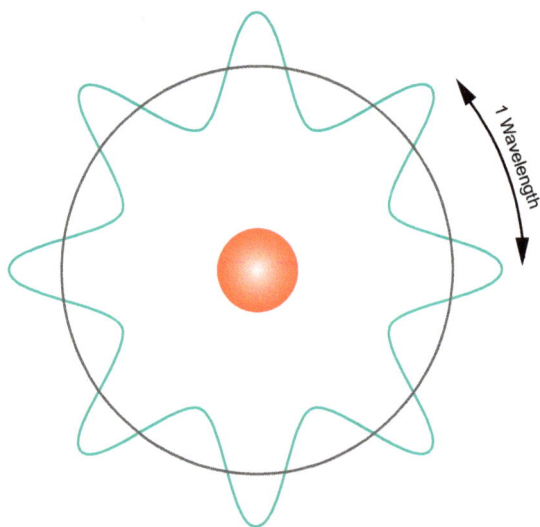

FIGURE 14: De Broglie's wave model for the electron. In this case the electron occupies an orbit with a circumference of exactly 8 wavelengths.

Now an atom has many possible electron orbits and de Broglie associated each of these with a different number of complete wavelengths, beginning with 1, then 2 and so on. The letter n is normally used to depict this series of whole numbers and the wavelength is normally represented by the Greek letter λ. So the equation may now be written as

$2\pi r = n\lambda$ where n represents any whole number.

This equation is starting to look a little like the series of values that Bohr had calculated by other means for the possible angular momentums of the electron that was discussed previously. In fact, De Broglie's equation can be manipulated and shown to be exactly the same as the series of values developed by Bohr. This is shown in Appendix 3.

It should be remembered that de Broglie's depiction of the atom as shown in Figure 14 is totally fictitious like all other attempts to draw an atom. It is simply a geometrical construction devised by de Broglie to illustrate his idea that electrons could be described as wave-like entities. As such, it worked and gave credible answers but the examiners of his thesis did not like the idea. Even though his mathematics could not be faulted, his examiners looked upon his work more as a trick of mathematics than reflecting anything real about the structure and behaviour of matter. They wanted experimental proof that matter could behave like waves not realising that experiments had already been conducted that proved just that.

De Broglie suggested performing the double slit experiment using electrons rather than light. If electrons exhibited a wave-like character then they too should exhibit a diffraction pattern on the screen similar to the light and dark stripes created by light. Such experiments have subsequently been done and electrons do behave like waves interfering with each other so that the electrons only arrive at certain places on the screen and not others. But, because of technical difficulties, these experiments were well in the future and as such could not help de Broglie.

Out of frustration, de Broglie's supervisor showed his student's work to Einstein who immediately saw the significance of what de Broglie was saying. Matter waves could most certainly be real Einstein wrote to Born at the university at Göttingen and, in turn, Born discussed the idea with experimental physicists at the institution who informed him that some recent experiments involving the diffraction of electrons off crystals might be explained by this new wave theory.

Even with this experimental evidence, the scientific community was reluctant to accept de Broglie's theory but experiments in 1927 finally gave his work the nod. Two independent groups showed that indeed electrons did behave in a wave-like fashion. One of these groups was led by George Thomson (1892 – 1975) the son of J. J. Thomson who had discovered the electron in the first place. So, we now have a very interesting historical situation. J. J. Thomson receives the Nobel Prize in 1906 for discovering the electron and showing it to be a particle. Then in 1937, his son George Thomson receives the Nobel Prize for demonstrating the electron to be a wave!

Both descriptions of the electron are correct. In fact, one of the most important implications of quantum mechanics is just that. All objects no matter how big or small have both particle and wavelike characteristics. It is not so much that a particle sometimes behaves like a particle and at other times behaves like a wave. It is behaving like both at the same time. We just cannot think of a simple analogy which describes such a possibility because simply there appears to be no analogy.

All this leads to an obvious question. If all objects behave in a wave-like manner why don't we sometimes see tennis balls behaving like waves? They do behave like waves, all the time, but the effect is so small it is insignificant and we do not notice it. On the human scale of things, large objects tend to behave more like particles while very small things like atoms and electrons behave more like waves. That is why it took so long to discover the implications of quantum mechanics. In Newton's day, large objects were all that could be observed so the wave-like nature of matter went unnoticed. Only when technology became available to pry into the atom did this strange new wave-like world manifest itself.

All objects obey the equations we have discussed linking particle-like properties with wave-like properties. The equations which work for electrons also work for tennis balls so that we can apply the equation linking the particle-like property of momentum p and the wave-like property of frequency f with the equation $p = hf / c$.

Now, compared to tiny electrons, tennis balls tend to have a very large momentum p which, using this equation where both h and c are constant, results in a very high value for the frequency f. It is this extremely high value for frequency which ensures the tennis ball in everyday situations behaves mostly like a particle. Electrons with much smaller momentum also have much smaller frequencies and it is these relatively low frequencies which contribute to the electron displaying wave-like characteristics.

To explain this contrast more fully, we need to look at the relationship between the frequency of a wave and its wavelength. Frequency and wavelength are related inversely to one another meaning that objects with high frequency can be said to have a very small or short wavelength and vice versa. So, tennis balls with a high frequency have a very small wavelength whereas electrons have a long wavelength.

The simple equation relating frequency with wavelength is $\lambda = c / v$, where c again is the speed of light, so the above equation for momentum can be re-written to replace frequency with wavelength, giving $p = h / \lambda$. These two equations ($p = hf / c$ and $p = h / \lambda$) relating particle-like momentum p with wave-like characteristics of frequency f or wavelength λ are therefore equivalent.

Now the reason that items with a long wavelength, such as an electron, clearly exhibit their wave-like character is because their wavelength is so long it is much larger than the width of each slit in the double slit experiment. As such, electrons will diffract through these slits—a necessary criterion for the experiment to produce an interference pattern on the screen. This is the criterion for exhibiting wave characteristics. Tennis balls do not behave like waves when projected towards double slits because their associated wavelength is much smaller than the width of each slit. As such, very little diffraction occurs and tennis balls are seen to behave like particles. Essentially, they go straight through the slits and impact the screen in very predictable straight trajectories following the rules of classical physics. This situation is shown in Figure 15a. Figure 15b shows the case where the wavelength of the particles is much larger than the width of the slits. As such, the expected diffraction pattern is formed on the screen due to interference between the two paths of wave-like

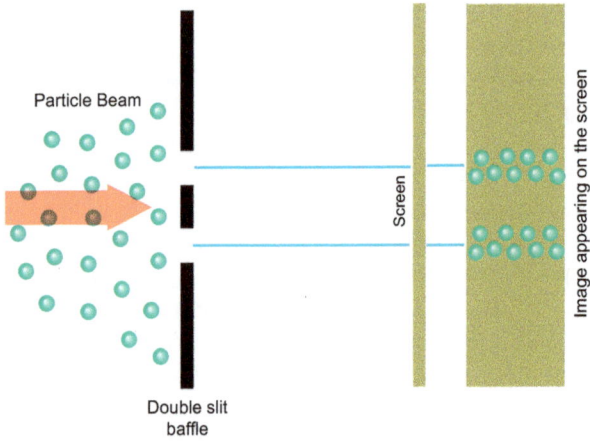

a. The expected classical situation where particles travel straight through the slits and arrive at the screen without interaction.

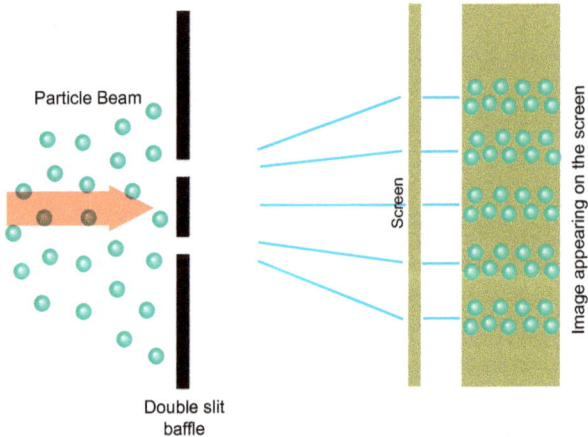

b. What actually happens in the quantum world.

FIGURE 15a and 15b: Particles passing through double slits produce differing interference patterns depending on their associated wavelength and width of the slits. According to quantum theory, the particles travel and interact as if they were waves and arrive at the screen as particles.

particles which, enroute, interfere as waves but arrive at the screen as particles.

It is still very hard to perform the double slit experiment with electrons because, compared to photons of light, they are relatively big particles. The width of each slit needs to be about the size of the distance between atoms in a solid crystal and this is very tiny. The experiment is much more easily performed using light. Photons are far smaller than electrons—as we saw before, their mass if they could be at rest would be zero. As such, the wavelength of light is much longer than the wavelength of electrons and so it is quite easy to perform the double slit experiment with light in any high school laboratory.

One well known application of de Broglie's theory today is the electron microscope which is used to look at very small objects which normal microscopes cannot resolve. An important factor in the function of all microscopes is the wavelength of light used. It just so happens that in order to resolve more detail, shorter wavelength light must be used. So violet light is better than red light for resolving fine detail. Better still, microscopes using gamma rays can resolve even more detail. Gamma rays are just about the shortest possible wavelength of light available and so with gamma rays, the conventional optical microscope reaches its limit of resolution.

Thanks to de Broglie, quantum mechanics tells us that other particles have wave properties and in fact the heavier the object and therefore the greater the momentum, the smaller the wavelength of that object. Could we construct a microscope which used a beam of tennis balls to illuminate or interrogate the subject? Unlikely, but electrons might be used instead of tennis balls. How does their wavelength compare with the wavelength of light used in normal optical microscopes? If it is shorter then the resolving power of microscopes could be improved.

Electrons are indeed much heavier than the photon particles of light and so have a much larger momentum p. Observing the equation $p = h/\lambda$, larger values of p mean smaller values of λ, h being a constant. The electron has a much smaller wavelength than any colour of light and even smaller than that of gamma rays, so electrons would be very appropriate for use in a microscope.

You might well ask how a beam of light could be replaced by a beam of electrons—they are very different things. An electron microscope looks quite different to an optical microscope. Optical microscopes have glass lenses to bend and focus the beams of light. Electrons are negatively charged particles so electron microscopes can utilize electric and magnetic fields in place of the glass lenses to focus the electron beam.

Beyond Reality

With de Broglie's suggestion that all matter had a wave-like character, the quantum world moved further away from reality. It was impossible to conceive of matter behaving partly like a solid particle and partly like a wave and any attempt to find or develop a metaphor from everyday experiences failed dismally. Gradually it was accepted that indeed trying to describe what atoms and electrons looked like was a futile exercise which would potentially even hamper further progress. So began the era when quantum mechanics finally parted with the methods of the old classical theory. There was now no longer any need to justify or attempt to explain the behaviour of the physical world in realistic or logical terms.

Up to this point quantum mechanics still contained a lot of classical baggage, at least in the attempts to explain the findings. Bohr described the atom as a miniature solar system with planet-like electrons with angular momentum circling a central nucleus. Later, the atom became a nucleus surrounded by clouds of electrons moving randomly in onion-shell rings. Then de Broglie described an electron as a wave which curved around the nucleus joining its tail and head. All these images served a purpose in so far as they allowed the development of a full mathematical description but none of them alone described everything and when taken together, they were often contradictory.

Progress began to be made by simply suggesting ideas, no matter how wild or classically non-sensical. If the idea explained part of the developing theory then it was accepted. Even so, researchers still found it difficult to become completely detached from the classical world and every new idea somehow needed to be explained in familiar terms but now nobody believed that these familiar metaphors really had anything to do with reality anymore. They were used simply to illustrate a point.

In building up a model of the atom it was ultimately found that several series of quantum numbers were needed to explain different aspects of the atom. It seemed that every aspect of the atom could be described using discrete quantised properties. Electrons could be found at various discrete energy levels and this required a series of quantum numbers to describe them as we have seen. Other numbers described other properties of the electron orbit.

In 1924, the American physicist Wolfgang Pauli (1900 – 1958) found the need to introduce a series of quantum numbers to describe something strange about the spectra emitted by the elements. The spectral lines came in pairs. Pauli interpreted this to mean that two electrons and no more could occupy the same energy level at one time. Pauli attached a quantum number to these electrons assigning a

Albert Einstein (l) and Wolfgang Pauli (r) talking and viewing papers at Leiden, autumn 1926.

(Photograph by Paul Ehrenfest, courtesy of AIP Emilio Segrè Visual Archives)

different value to each. This was in fact the fourth quantum number to be required to describe the atom at this stage although in this discussion we have only looked at one other, the principal quantum number n which described each energy level.

So was born the Pauli Exclusion Principle which stated that no two electrons within an atom could have the same set of four quantum numbers. The state of every electron must be described by a unique set of quantum numbers. Pauli was in his early twenties and had the privilege of working as Max Born's assistant at Göttingen as well as spending some time with the great master, Bohr, in Copenhagen. Pauli was a friend of Heisenberg, who enters the quantum story shortly, and like Heisenberg was liberally afflicted with the arrogance of youth. For his work, Pauli received the Nobel Prize in 1945.

So what did Pauli's new quantum number describe? Remember that by the mid-1920s researchers had realized the futility of attempting to explain or tie new quantum characteristics to familiar ideas—ideas such as electrons actually orbiting the nucleus were now considered rubbish. Even so it was difficult not to try to visualize new atomic properties in terms of familiar classical models.

So it was with Pauli's new quantum number series. It was finally given the name electron spin and had a value of either up or down as shown in Figure 16. It might seem strange that the direction of the spin of an object can be described as either up or down. Surely clockwise or anti-clockwise would be more appropriate. These terms can be ambiguous depending on which way the object is viewed. Spinning involves a rotation about an axis through the centre of an object so it is this axis which describes the spin and by convention a certain direction along this axis, either up or down, is adopted to indicate the direction of the spin.

The naming of quantum spin was yet another attempt to describe a quantum property in terms of something classically familiar. The spin of an electron was ultimately described using analogous mathematics as a spinning tennis ball. However, this quantity was discrete too! All electrons appear to be spinning at the same speed in either a clockwise or anti-clockwise direction. The electron is such a strange particle that two revolutions are required to get back to where it started as shown in Figure 17. Obviously, no macroscopic world object ever does this. Again, it became clear that there was

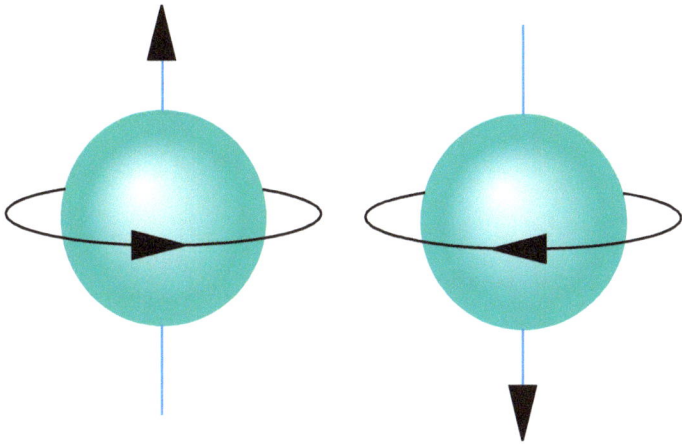

FIGURE 16: The convention of describing electron spin as either up or down.

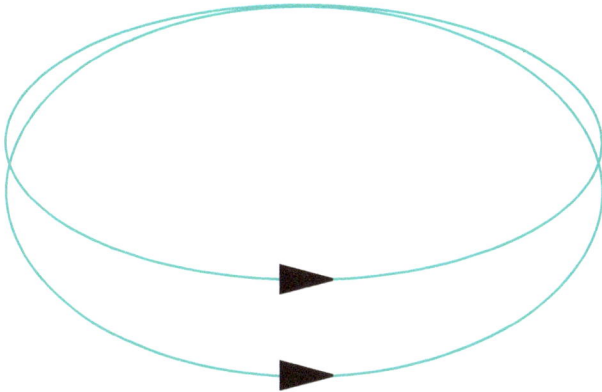

FIGURE 17: Electron spin requires the electron to orbit twice in order to get back to its starting position.

little point trying to represent the makeup of the atom strictly in terms of everyday metaphors. Clearly the microscopic atomic world was full of strange behaviours the macroscopic world was totally unaware of and could not comprehend. We continue to use the term spin because the quantum mathematics is almost identical to classical angular momentum, the only difference being that we don't observe the spinning! We can only probe spin by its quantised magnetic moment.

Quantum Mathematics

By the mid-1920s, quantum mechanics looked like a confused array of seemingly unrelated bits and pieces of a theory loosely and rather tenuously tied together by a group of simplistic numbers referred to as the quantum numbers. In Göttingen, Born could see no way ahead. Einstein was expressing doubts about the role of probability in the theory—for example and amongst other things, how it related to the timing of an electron which seemed to decide of its own accord when to drop to a lower energy state. Einstein believed there was some underlying theory strongly embedded in the old classical physics which would explain this timing.

Of the grand masters, only Bohr expressed some hope although he was becoming depressed about the contradictions which were starting to appear in his model of the atom. Bohr's approach had been to quickly accept radical ideas if they suited the observations and then move on. This approach helped him devise his brilliant model. Meticulous mathematicians like Born were concerned there was no comprehensive theory which brought everything together and in fact in 1925 he considered it would be many years before such a theory would be thrashed out, if ever. But then he did not count on the coming brilliance of a new and younger generation to which Heisenberg belonged.

Werner Heisenberg (1901 – 1976) studied under Arnold Sommerfeld (1868 – 1951) at the University of Munich. Sommerfeld gave the young German the job of finding a mathematical solution to the problem Pauli had looked at involving the spin of electrons. Much to Sommerfeld's amazement, Heisenberg came up with a very simple solution within a matter of weeks. Heisenberg suggested a

78

Niels Bohr (l) and Werner Heisenberg (r) outside in the snow., circa 1932. Bohr seems to be working on something important or is it just a ski guide!

(Max Planck Institute, courtesy AIP Emilio Segrè Visual Archives)

series of quantum numbers with half-integer values, namely 1/2, 3/2, 5/2 . . . Sommerfeld quickly scuttled the idea because, up until this time, quantum numbers had always been thought of as whole numbers. It took the young mind of Heisenberg, not steeped in tradition, to suggest a radical idea which later turned out to be totally correct. Unfortunately, he missed out on the accolades because of Sommerfeld's conservatism. The credit went to others who were not so restrained in their enthusiasm to publish.

The incident did not hold back Heisenberg for long. During stints as Born's assistant at Göttingen and also with Bohr in Denmark, Heisenberg went on to develop a comprehensive theory of quantum mechanics by adopting the attitude which was now widespread—that there was no longer any need to tie quantum physics to old classical origins. Working intensively and often through the night, Heisenberg developed a radical theory using, what were at the time, strange mathematical methods known as matrices, with which he was only vaguely familiar—and that required that a x b does not give the same result as b x a. Gradually the observed phenomena fell into place in his theory and, in fact as it developed, his theory began to predict the varied and diverse aspects of the entire quantum theory to that date.

The story goes that in May 1925 Heisenberg went for a rest break to the rocky island of Heligoland in the North Sea off the coast of Germany to recuperate from a bad bout of hay fever. Rested, Heisenberg was able to use this quiet time to reflect on his research and soon came up with the first comprehensive mathematical treatise of quantum physics. Hesitantly he presented his calculations to Born who, after some trepidation, realized Heisenberg was on to something. Born recalled a lecture he once attended on basic matrix theory and realized this would help formalize the work that Heisenberg had done. Born and Heisenberg, together with the more mathematically skilled physicist Pascual Jordan (1902 – 1980), went on to more fully expand Heisenberg's mathematical theory which came to be known as matrix mechanics.

The beauty of the theory was that it explained all aspects of mechanics, not just the new quantum workings of the atom. The classical mechanics of Newton were also encompassed within the theory and Newton's laws were now seen for the first time as merely a special case of the more general theory when it was applied to larger everyday situations. This was similar to the situation that occurred with Newton's laws in the wake of Einstein's theory of relativity. Einstein's theory was able to describe all of Newton's classical theory as well as the revelations of the new relativity theory.

A matrix is a multi-dimensional mathematical object, the elements of which can be other mathematical objects, for example numbers, complex numbers, operators or even other matrices! Matrices are different from numbers because they are described by a different algebra. The more familiar numbers such as 3, 4, 5 describe only one thing, that is, a single value of a certain quantity. A matrix can be made up of several numbers each describing a different aspect of some phenomenon. A typical simple matrix might look like that shown in Figure 18 with each of the four numbers describing

$$\begin{pmatrix} 1 & 3 \\ 5 & 7 \end{pmatrix}$$

FIGURE 18: A matrix is made up of several objects (in this case, ordinary numbers) each describing the value of a different aspect of some phenomenon. The numbers could be written in many different ways, such as simply (1,3,5,7), but conventionally an array format has been chosen with each position in the array representing a particular parameter and the number at that position in the array giving a numerical value to that parameter. Mathematically, the array or matrix format is the most useful. There are rules for adding and multiplying matrices. This example is a 2 x 2 matrix as it has two rows and two columns.

a different characteristic of something. The numbers could be written in many different ways, such as simply (1,3,5,7) but more usefully an array format was developed with each position in the array representing a particular parameter and the number at that position in the array giving a numerical value to that parameter. An array format is more powerful mathematically than a simple linear format. Matrix mathematics has been found useful in many different fields, not only quantum mechanics. It can make the mathematics more manageable in comparison to an equivalent set of equations. Rules relating to adding or multiplying these matrix numbers have been developed and often they are not much different to the rules for ordinary numbers. Sometimes, however, the rules are quite distinct. We all accept that 2 multiplied by 3 is the same as multiplying 3 by 2 but with matrices this is not the case. The order in which matrices are multiplied is important in much the same way as the order in which ordinary numbers are divided is important. Dividing 3 by 2 is not the same as dividing 2 by 3.

The new matrix mechanics showed that, in the quantum world, multiplying the momentum p by the position x of a particle produced a different value depending on the order of the terms (10) and in fact the difference of these two terms was $px - xp = h / 2\pi i$, where h is again Planck's constant and i represents the square root of minus one. Now the square root of minus one turns up quite a bit in quantum mechanics and is usually represented in equations by the letter *i*. The square root of minus one is represented by *i* because there is no real number to represent it. This is explained more fully in Appendix 4.

The square root of -1 is neither 1 nor -1. Try the reverse—try squaring either 1 or -1. In either case you cannot get back to -1. 1 squared gives you 1 while -1 squared also gives an answer of 1 because a negative number multiplied by a negative number gives you a positive number, so a real number solution to the square root of -1 is not possible. Hence the need for it to be represented by *i*.

At the time, matrix mechanics was as difficult to read as its implications were to comprehend so it was not surprising that few researchers grasped the significance of Heisenberg's mathematics. There was an exception. The English physicist Paul Dirac (1902 – 1984) was the same age as Heisenberg and as chance would have it went to a lecture given by the young visiting German at Cambridge prior to Heisenberg again, with Born and Jordan, presenting their combined paper on matrix mechanics. The year was 1925.

Heisenberg's first lecture was of course on quantum mechanics but he did not talk about his ideas on matrix mechanics because he was still unsure of himself—he was only 24 years old. He did however later send a draft copy of his first paper on matrix mechanics to Dirac's supervisor whom he had met when at Cambridge. This was subsequently passed to Dirac making him one of the first people to read of Heisenberg's new theory.

Dirac was well versed in mathematics, having first studied engineering before changing disciplines and undertaking a degree in mathematics. He first encountered quantum physics when he arrived at Cambridge to study for his doctorate and had no trouble assimilating the ideas presented by Heisenberg. In particular, the importance of the order in which matrix terms were multiplied struck a chord with him—the so-called non-commutativity of matrix multiplication. Dirac recognized this in the formulations of the Irish mathematician William Hamilton (1805 – 1865) who almost a century before had developed a new and rigorous branch of mathematics which explained most of the Newtonian classical mechanics—the Hamiltonian theory.

Hamiltonian theory (11) represented a whole new way of looking at Newton's laws of motion which Newton developed in the late 1600s.

Much of Newton's theory involves the acceleration of objects, or more verbosely, the rate of change of the rate of change of position of an object—the so-called second order equations. By concentrating on the parameters of position and momentum, rather than velocity, Hamilton was able to describe a theory using first order equations which only involved rates of change, rather than rates of change of rates of change. Hamilton expressed the total energy of a system in terms of its position x and momentum p in one single quantity which he called the Hamiltonian function H. Using this function, it was possible to formulate two independent equations for the position and momentum of the system. It was considered a unique and elegant approach to the field of mechanics and would in turn, as we are about to see, have even more far reaching applications in the broader field of mechanics than Hamilton could ever have imagined. The Hamiltonian equations were found to be true for any classical system and not only Newton's basic laws of motion. As such, they applied to Maxwell's theory of electromagnetic radiation as well as Einstein's relativity.

Starting with the formulae of Hamilton, Dirac was able to develop a mathematics of his own which he termed quantum algebra and which gave the same results that Heisenberg had arrived at using his matrix methods. Within four months of reading Heisenberg's draft paper, Dirac's theory was also in print. It is not known whether Heisenberg was shocked at the quick response from his contemporary but upon reading Dirac's paper Heisenberg sent a congratulatory reply commenting that he felt Dirac's work had presented his own findings in a more concentrated and beautiful form (12).

The mid 1920s were indeed a fertile era for physics. Those with the mind, appropriate attitude and diligence were able to quickly capitalize on each successive breakthrough and push ahead quickly with new ideas and concepts. They worked hard and obsessively—sometimes around the clock. Their work was everything. Others fell by the wayside or could only look on from the sidelines as these breathtaking advances were made. These were the halcyon days for physics in the modern era.

The Double Slit Revisited

Some of the strange mathematics of quantum physics can be glimpsed at by once again looking at the double slit experiment (12) and the interference pattern produced when two light waves interact, but note that the following discussion can only best be described as superficial as much is left out or not explained. For those with little interest in mathematics, this section could be completely omitted.

Remember that, earlier, a quite adequate explanation of the interference of two waves was given using the wave model for light. Using water waves, when two wave crests met, they reinforced each other producing an even higher wave whereas, when a crest and a trough coincided, they cancelled out to produce flat water. Using light in the double slit experiment, these interactions caused either a bright or dark region on the screen.

As we have seen, this wave interpretation represented a dead-end street to understanding all aspects of light and so was born quantum physics. With it, alas, also came the complex mathematical formulations developed by Heisenberg, Dirac and others. The following discussion may gloss over much of the theory and avoid the complex mathematics but it at least describes the essence of quantum mechanics.

Two slits offer light two possible paths along which to travel and each of these paths can be assigned a specific probability of occurring, namely p and q. Note that we have already used p to describe the momentum of a particle and this is its normal use in physics. It just so happens that the word probability starts with the letter p so I have also used p to describe probability just for this exercise. Now if these are the only two paths, then $p + q$ must equal 1. This is the same as saying if these are the only two paths then there is a 50% chance of a photon of light using slit 1 and a 50% chance of a photon using slit 2. In other words, a 100% chance (50% plus 50%) that the photon will use one of the two slits because there is no other path for it to take. In probability theory, decimal values instead of percentages are sometimes used. Then, fractions of 1 indicate probabilities, so 50% equates to a probability of 0.5 and $p + q$ equals 0.5 + 0.5 or, in other words, 1.

Now in quantum theory p and q are considered to not quite represent probability but rather something more arcane called probability amplitudes and these amplitudes are not simply real numbers like 0.5 but rather a combination of real numbers and imaginary numbers containing that strange number i representing the square root of minus 1 which was explained earlier. Remember that the square root of minus one is represented by i because there is no real number to represent it. In quantum-speak one might say there is an amplitude p of a photon going through slit 1 and an amplitude q of it going through slit 2.

Amplitudes are not probabilities but they are remotely related. In quantum theory, to get the real value for the probability of any event happening, the amplitude must be squared—that is what the mathematics says. So the probability of amplitude p occurring is p^2.

In situations where there are two possibilities, such as with the double slit, quantum rules say that the amplitude (not the probability) of a photon arriving at the screen is the sum of the amplitudes of the two possibilities, so this total amplitude becomes

p *(of alternative 1)* $+ q$ *(of alternative 2)*.

Now, if one slit is closed, one of these terms drops out and the probability of a photon arriving at the screen becomes p^2 or q^2 depending on which slit is left open. Remember the probability is the square of the amplitude. If both slits remain open we have to then square the sum of the amplitudes *(p + q)*, that is *(p + q)²* and since these amplitudes may contain imaginary numbers, the answer is not simply $p^2 + q^2 + 2pq$ as it would be using basic algebra with real numbers but something more complex, namely

$p^2 + q^2 + 2pq \cos \theta$

which contains an extra strange-looking term incorporating *cos θ*. *θ* represents an angle in a strange mathematical world where real and imaginary numbers are combined to form what mathematicians call complex numbers. *Cos θ*, or the cosine of *θ*, may be a little easier to understand. *Cos θ* can only have values ranging from -1 to +1. You

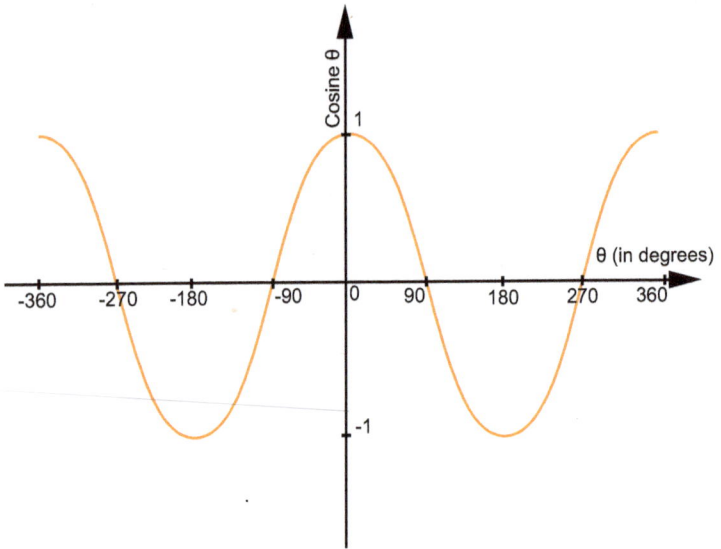

FIGURE 19: A graphical plot showing the classic shape of the cosine wave.

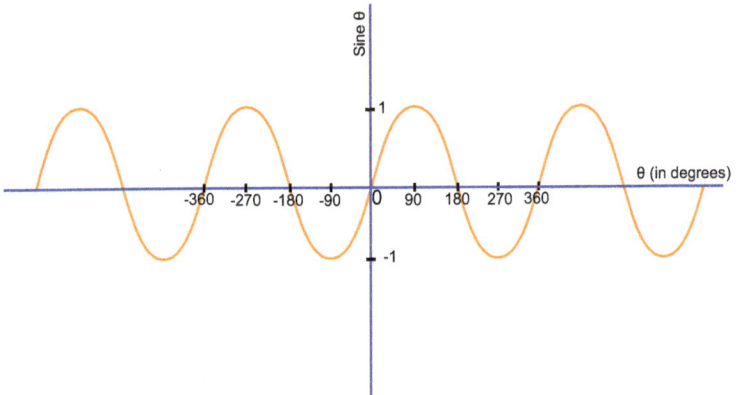

FIGURE 20: A graphical plot showing the classic shape of the sine wave. The sine wave (and cosine wave) extends to plus and minus infinity values of θ with monotonous regularity.

can check this out on your calculator using the angles 0, 90, 180 and 270 degrees if the calculator has cosine values.

You may be familiar with a sine wave. Cosine waves are the same shape and one is shown in Figure 19. A sine wave is shown in Figure 20. Sine and cosine waves have the same shape. Their only difference is in the phase between them. A sine wave starts at, say, time zero with a value of zero whereas a cosine wave begins with a value of 1 at time zero. Sine and cosine waves depict the classical wave shape and these curves are involved in many physical and mathematical situations involving cyclical processes—classical wave theory included. So, it is no surprise that they appear again in quantum theory which does to some extent involve wave motion even though it may only be metaphorical. It is important to realize however that the cosine wave used here does not in any way relate to any real wave in nature. It is simply a mathematical element of the theory but it is of course difficult not to think of it being in some way related to reality.

It is the term *2pq cos θ* which provides the quantum interference between the two beams of photons and so produces the light and dark patterns on the screen. If $\theta = 0$ degrees then *cos θ* is 1 and the term *2pq cos θ* is a positive *2pq*. The overall probability, $p^2 + q^2 + 2pq$, of the two waves reinforcing each other is therefore high. If $\theta = 180$ degrees then *cos θ* equals -1 and the term *2pq cos θ* becomes negative *2pq* so making the overall probability of the waves reinforcing low.

For the double slit experiment in particular, with both slits open, it can be shown that the bright regions on the screen occur when $p = q$ (since the slits are identical) and *cos θ* equals 1. Then the full equation for the combined probability discussed previously becomes

$$p^2 + p^2 + 2p^2 \text{ (replacing } q \text{ with } p \text{ since they are equal)}$$

which equals $4 p^2$ meaning the probability (which can now be equated to light intensity at this point on the screen) is four times the intensity at this point of only one slit being open. Why is it not just double? Astounding as this may seem, this fully agrees with experimental findings.

For the dark regions on the screen, where two waves cancel each other out, it can be shown that $p = -q$ and $cos\ \theta$ equals -1. Then the full equation for the combined probability becomes

$$p^2 + p^2 - 2p^2$$

which equals zero, again agreeing with what is observed.

The above analysis illustrates just how far from reality quantum thinking had become and how powerful the mathematics had become. An analysis of the real world no longer seemed relevant as long as the correct answer was found. Using classical theory, we could arrive at the same conclusion by simply using the water wave analogy and adding or subtracting the wave heights or amplitudes. That was relatively easy to understand and related to something in the real world.

In quantum theory we are manipulating weird numbers—real and imaginary to achieve the same thing. The mathematics seems like a fudge but this was precisely the point. Nature behaved in ways we could hardly imagine. If the equations gave the correct answer then these equations were the best description available of nature at that time. It was as simple as that.

One must surely ask whether these quantum equations do actually contain a real picture of nature. Many physicists said it didn't matter and were quite happy to simply calculate the probabilities of events and leave it at that. Others believed it was intrinsically impossible to obtain a real picture of nature's workings. Physics had gone beyond that point and besides, what is meant by real anyway? When looked at more closely it seemed that even the old classical physics was built on a series of metaphors which weren't necessarily real. All that quantum physics did was sweep away the pretence.

In the middle of the 1920s there was one prominent physicist who saw no reason to break with the past and who indeed was determined to reconcile the new quantum physics with the old classical view and once and for all curtail the radical movement towards a probability model of nature where indeterminacy ruled supreme. This person was the Austrian physicist Erwin Schrödinger (1887 – 1961) who shared the Nobel Prize with Paul Dirac in 1933. In the mid-1920s,

Left to right, Heisenberg's mother, Schrödinger's wife, Dirac's mother, Paul Dirac, Werner Heisenberg and Erwin Schrödinger at Stockholm train station, Sweden, 1933. (Max Planck Institute, courtesy AIP Emilio Segrè Visual Archives)

Schrödinger was nearly forty—almost a generation older than Heisenberg, Dirac and Pauli at a time when it seemed youth was the necessary prerequisite for quantum discoveries. Schrödinger was professor of physics at the University of Zurich in Switzerland and this conservative institute was not to be the place for radical thought. Schrödinger integrated quantum mechanics into an old-fashioned theory which made sense and was more easily understood. It produced the correct answers and seemed in fact to be reconciled with classical physics. It provided the new discipline of quantum physics with a very useful tool and for this reason is still used to this day. Schrödinger's theory is worth investigating.

Schrödinger's Wave Theory

A lot was happening in the mid-1920s. In 1923, de Broglie had identified the wave nature of particles as a counterbalance to the particle description of waves. Then, in 1925, both Heisenberg and then Dirac gave their separately-derived but equivalent mathematical descriptions of the quantum world. These two descriptions were quickly accepted but remember that it took a few years before de Broglie's wave view of matter became widely known. It was this wave aspect of matter developed by de Broglie that Schrödinger picked up and carried into his own theory which became known as wave mechanics to distinguish it from Heisenberg's matrix mechanics and Dirac's quantum algebra.

Starting with de Broglie's idea of electron waves encircling the nucleus of the atom, Schrödinger developed a wave theory to describe the electron energy levels of the atom. To incorporate the idea of quantum jumps, he again used de Broglie's idea that the orbits of the electron waves encircling the nucleus must be exactly a whole number of wavelengths in circumference. Fractions of a wavelength were not allowed. Schrödinger published this fine piece of work in 1926, a year after the release of the theories of Heisenberg and Dirac.

Schrödinger's theory used waves which existed in the real world of classical physics and with which everyone was familiar. His theory seemed to be developed from reality whereas Heisenberg's and Dirac's, although giving the right answers, were not. As de Broglie had pointed out in his wave theory of matter, his concentric waves encircling the nucleus were only a geometrical construction which gave the correct answer. No one believed that electrons were waves encircling the nucleus. In fact, no one knew what electrons were. So, was Schrödinger's theory closer to reality than the other two? Amazingly, all three theories were ultimately shown to be the same although Niels Bohr, working in Copenhagen, showed that Schrödinger's wave theory was not all that good at describing some quantum phenomena.

Of the three theories, Schrödinger's was more quickly accepted, especially at first because it did seem to relate more closely to real things through its derivation from wave theory—and waves had been accepted as part of reality for a long time. This connection however was only an illusion. Again, the use of waves in the Schrödinger theory was merely a form of geometrical construction thinly disguised as reality. Many have argued that this continued attachment to describing the world of electrons through a wave theory has done more harm than good in furthering an understanding of quantum mechanics. It might have been far better, it was argued, had a clean break with the past been made, so declaring there was no possible physical model or complete metaphor which could possibly describe the workings of the atom.

The true worth of Schrödinger's theory centres on the term he called the wave function (14) which in physics is usually represented by the symbol Ψ. Generally, Ψ represents a mathematical equation. Sometimes this wave function is called the state function in order to remove any apparent linkage to a wave theory of reality. Mostly, state function is the term used here.

To explain what is meant by this function, imagine a very simple situation involving a single particle whose state we wish to describe at any moment. This is called the particle's quantum state. A point-like particle's state is just its position in space at any time and in order to predict what that particle might do next it would help to know if it is moving or not. In other words, we also need to know its speed or velocity or even its momentum. Then we could predict its position at a later time. In fact, we could trace out its path as it moved through space and this is basically what is meant by the state function of that particle. Again, it is more useful in quantum mechanics to talk about the momentum of a particle rather than simply its velocity. Velocity is tied up in momentum anyway along with the mass of the particle so there is no problem in shifting from velocity to momentum in the state function equation.

In quantum mechanical language, the state function Ψ can be thought of as every possible position that the particle might have available to it, so again we can see probability creeping back into the argument. There is plenty of this in quantum mechanics. For any position, denoted by say x, the state function Ψ will have a certain value which traditionally is written as $\Psi(x)$, meaning the value of Ψ at x. In quantum mechanical language, $\Psi(x)$ is the amplitude for the particle at position x and, as was done previously for the amplitudes of the alternative pathways in the double slit experiment, we need to square the absolute value of the amplitude to obtain the probability of finding the particle actually at position x. So, the chances of finding the particle at position x is $|\Psi(x)|^2$. The vertical brackets imply the absolute value—the value of $\Psi(x)$ without any sign, + or -, has first been taken before it is squared.

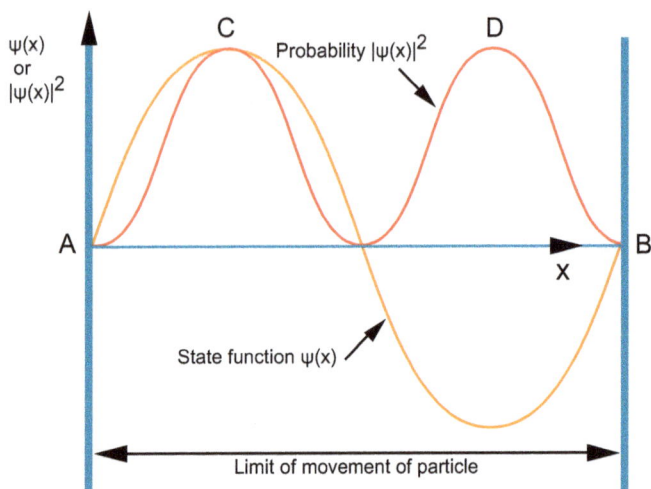

FIGURE 21: Plot of a sine wave (orange) as a one-dimensional state function ψ(x) together with |ψ(x)|², the probability of finding a particle at any position x (red).

Now what does this mean? Well, $\Psi(x)$ is simply an equation which could be plotted on a graph. It is however a complicated equation which usually involves real and imaginary numbers, but a simple analogy can be drawn using a very basic equation as shown in Figure 21. I have chosen a simple sine wave to represent an example of a state function, so $\Psi(x) = sin\ x$. If we assume the particle can only move back and forth in one direction, say x, rather than through real three-dimensional space then we could plot this position along the x axis and the value for the state function $\Psi(x)$ along the vertical axis.

If we were to compute the probability density $|\Psi(x)|^2$, this would give us a probability distribution for the possible locations we would likely find the particle as depicted in Figure 21. Clearly, looking at the figure, there are two lobes where we are most likely to find the particle. As well, we are unlikely to find it in the centre or at the boundaries. The wave function $\Psi(x)$ gives us a sure way of computing the probability of the particle's location but what does it physically correspond to? To this day, this question is still hotly debated in the physics community.

Recapping, the value of $\Psi(x)$ for any value of x is the vertical distance from the x axis to the curve and the probability of finding the particle at this position is the square of this distance. So, it is possible to plot a line representing $|\Psi(x)|^2$, or the probability of finding the particle at any position, by calculating a value for $|\Psi(x)|^2$ at every point x and drawing a curve to link these points. This is how the two-lobed curve shown in Figure 21 is constructed. You can verify the shape of this probability curve by actually making some measurements of the state function $\Psi(x)$ from the graph and squaring them. Notice how the negative parts of the state function become positive probabilities because any negative number squared becomes positive. A negative number multiplied by a negative number results in a positive number. If this were not the case it might be quite difficult assigning a meaning to a negative probability.

Overall, it is easier to understand what is meant by the probability curve than it is by the curve representing the state function. For example, the state function $\Psi(x)$ drawn in Figure 21 might represent the fictitious situation of a particle in a box with the positions A and B representing the side walls. We are only concerned with the side walls because this is an imaginary one-dimensional box. The state function chosen (*sin x*) is also a very simple case and also completely fictitious since, as we shall see, real particles in boxes do not behave according to this state function.

The associated probability curve $|\Psi(x)|^2$ tells us that in this case we are most likely to find the particle at either of two places C or D since the probability curve is highest at these two points. We are unlikely to find the particle next to either wall as the curve representing $|\Psi(x)|^2$ drops to zero at the walls and we are unlikely to find it in the centre of the box as the curve does likewise there. Such are the wonderful workings of the Schrödinger state function. It is possible by delving heavily into mathematics to develop more complex state functions than the one described here, and these complicated equations actually do describe real situations.

The Principle of Complementarity

There seemed to be endless attempts to define nature as particles or nature as waves or a combination of both. In fact, no approach was totally satisfactory in explaining everything and hence the departure of both the theories of Dirac and Heisenberg into a totally abstract world. Only Schrödinger's theory attempted to hang on to the reality of waves.

What came out of the wave–particle dualistic nature of light has become known as the principle of complementarity and it was finally clarified by Bohr in 1928. It had been observed for some time that certain aspects of light could best be described using a wave model whereas other aspects needed light to consist of tiny particles. Further, if one description was chosen then that usually excluded the simultaneous use of the other description. There seemed to be no theory which seamlessly combined both waves and particles. Both wave and particle theories were needed. They formed a complement in order to obtain a full description of light.

The need for such a principle underlies the fact that understanding what light or electromagnetic radiation really is, is just too hard. There is nothing in our everyday experience like it from which we can draw a metaphor. Think about it. Can you invent a metaphor that simultaneously combines both waves and particles? We need several metaphors in order to build up the full picture. More than this, through these metaphors we are never allowed to view the full picture because using one metaphor, for example waves, precludes us from viewing any particle aspect, and vice versa. The two metaphors if used together are contradictory.

Max Born, in Göttingen, weighing up both sides of the theory, saw the practicalities in Schrödinger's approach but couldn't accept that his wave theory really represented particles composed of waves. He considered that electrons were real particles but they were particles whose position could never really be pin-pointed and the usefulness of Schrödinger's approach was to quantify a probability of finding an electron in a certain place. It wasn't possible to explicitly determine this place but instead only specify a probability of finding an electron there.

This is similar to driving a car across town to a certain destination, say half an hour away. You cannot precisely predict the moment you will arrive there because of all sorts of factors including traffic holdups or losing your way. These factors will make the outcome uncertain but you could probably estimate your time to within ten minutes.

There is however one major difference with this analogy. It would be possible to almost exactly predict your time of arrival if you knew all the facts. If you knew just how much traffic congestion there would be, if you made sure you knew the route exactly, then you could improve the preciseness of your estimated time of arrival quite considerably.

Not so with the electron. It seems that there is no way of improving the chances of finding an electron at a certain point simply by knowing more information about the electron's path. This path is essentially unknown and it seems like it can never be known. This was the big break with classical physics which had been built on a basis of total predictability—if you knew the speed of the planets in the solar system then you could precisely predict where they would appear in the night sky at any time. In quantum mechanics, the endpoints may be known but there is no information available as to how the particle got from the start to the finish.

What Schrödinger's state function really meant was that an atom's electron could be anywhere, even outside the atom in a different room of the laboratory. Of course, the probability attached to this scenario was extremely small, so small in fact that it would probably never happen. The new theories of quantum mechanics were making the world a very uncertain place.

Heisenberg's Uncertainty Principle

Is it possible to create a particle using waves? At first, this seems absurd. By their very nature, waves and particles seem to be opposites. Particles are well defined points occupying a certain position in space. Waves are nebulous and extend forever repeating themselves over and over again with no particular central point or focus which could be reasonably considered to represent the particle portion of

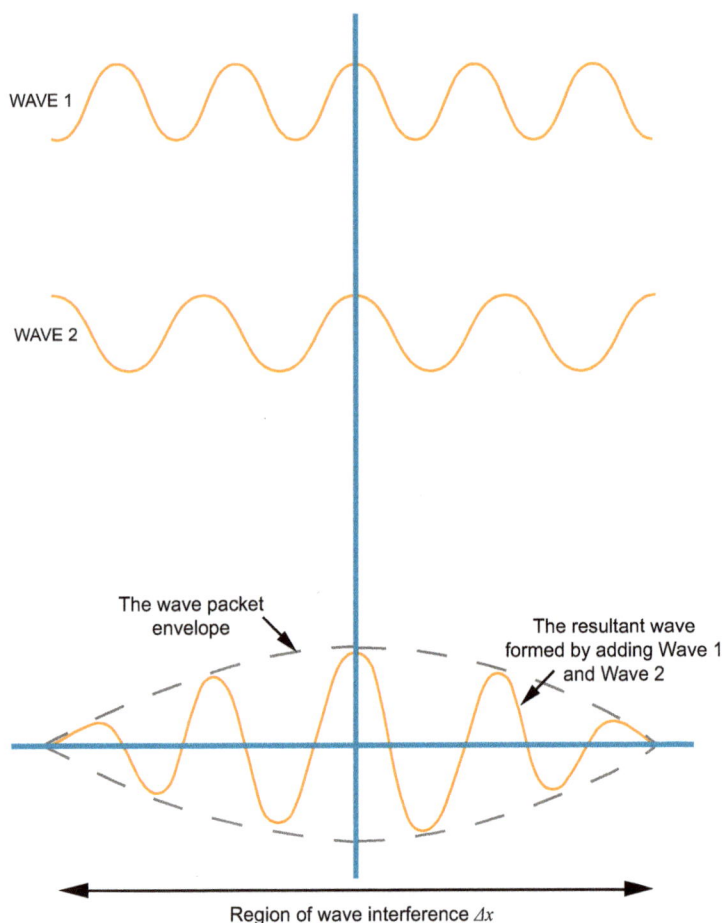

WAVE 1

WAVE 2

The wave packet envelope

The resultant wave formed by adding Wave 1 and Wave 2

Region of wave interference Δx

FIGURE 22: Adding several sine waves of different frequency (or wavelength) produces a more defined and localized shape. In this case, a wave envelope or packet, as it is called, is produced by adding two sine waves of slightly different frequency (Wave 1 and Wave 2) across the region of wave interference Δx. The resultant wave is the wave function $\psi(x)$.

the wave. Waves have no localised focus as shown by the simple example of a sine wave depicted previously in Figure 20.

If, however, a number of sine waves of differing frequencies (or wavelengths) interact with each other something interesting does happen. In this discussion I have chosen to use wave frequency but an argument using wavelength is just as valid. Remember an earlier discussion in which two water waves interacted in a wave pool. If two crests came together then suddenly a higher peak was created at that point. This certainly is a more localised focus. If we combine a number of waves differing in frequency over a range of frequencies the localized point becomes even more distinct as shown in Figure 22. These waves interfere over a limited distance to create a distinctive waveform over that limited distance.

The resultant shape is known as a wave packet and has been produced by waves with a range of frequencies (usually denoted by Δf) interfering with each other over a specific region of space (usually denoted by Δx). This resulting and reasonably localised wave packet represents a fuzzy particle. Fuzzy in so much as it is still not clearly defined—the overall envelope shape containing the wave packet still tapers off gradually at either end. If more waves over a larger range of frequencies Δf are used to create the wave packet then the cut-off at either end becomes sharper and the wave packet or particle becomes more defined.

Like so many of the analogies in quantum mechanics, this wave construction of a particle is not meant to relate to anything in reality. It is simply a mathematical construction to illustrate a point. Explaining the fuzziness of a particle's position using waves in this way does not vindicate the Schrödinger camp, which is that particles are really waves. The point to be made is that a fuzzy particle appearing over a region Δx can be described metaphorically using waves over a range of frequencies Δf.

If we want to make the particle less fuzzy, that is, make the shape of the wave packet more sharp-edged and hence more defined over a tighter region Δx, we would have to include additional waves over an even broader range of frequencies Δf. In other words, to pinpoint the particle more precisely (make Δx smaller) we need to increase Δf. Conversely, if we decrease the range of frequencies Δf down to a single frequency, so making Δf very small or in fact zero, then we are back to the case shown in Figure 20 where the wave is totally uniform with no confining region and Δx is now spread out to infinity. When we confine the frequency to just one value, the particle has no position at all—it is spread throughout the entire universe. Sensing the significance of this, Heisenberg seized the moment.

This, he realized, was the connection between the wave and the particle dualism of matter that had eluded everyone. The small region Δx was a particle-like characteristic whilst the frequency range Δf was obviously related to waves. The principle of complementarity precluded the simultaneous use of both the wave and particle characters of matter and now Heisenberg saw a way of quantifying this inverse relationship—less of one character implied more of the other or, in other words, a smaller ambiguity in the position Δx implied greater ambiguity in the frequency Δf.

Knowing that it was possible to relate frequency with momentum (using the equations that were previously derived in Appendix 3), Heisenberg was able replace the range of frequencies Δf with a range of momenta Δp and, after determining a value for the multiplication of these two quantities, came up with his famous uncertainty principle, namely

$$\Delta x \, \Delta p \geq h \, / \, 4\pi$$

which again features Planck's constant h and the universal constant π. (The use of the constant 4π in this equation is optional and depends on how the values of x and p are calculated but this in no way detracts from the profoundness of this equation. Likewise, other variables besides x and p have been used and 4π is sometimes written as 2π.)

The implications of this seemingly simple equation are profound and, in many ways, it embodies the central core or essence of quantum mechanics. It implies there is always a certain inability to determine measurements exactly and no matter how precise scientific instruments are or may become in the future they will never be able to eliminate this uncertainty. Uncertainty in measurement is a part of nature.

The principle of uncertainty applies to all situations whether they be tiny measurements on the scale of the atom or larger engineering measurements made in the design of a bridge or the trajectory of a tennis ball on centre court at Wimbledon. This may come as some surprise since there seems to be no doubt as to where tennis balls strike the court, at least these days with computers and video coverage. So, is Wimbledon exempt from Heisenberg's Principle?

Definitely not. It still holds on the tennis court just as it does everywhere else but its effects are very small and go totally unnoticed. Remember Heisenberg's uncertainty principle states that the product of the uncertainty in position and the uncertainty in momentum must be equal to or less than Planck's constant divided by 4π. Planck's constant is an extremely small number equal to about 6.6×10^{-34} and dividing it by 4π makes it even smaller. Now this scientific notation is quite easy to read when you realize that 1×10^{-5} means 0.00001. So, 6.6×10^{-34} is very, very, very small indeed. The -34 means move the decimal point in the number 6.6 thirty-four places to the left. Now the momentum of a tennis ball is very, very big compared to that of a photon or electron, so in Heisenberg's equation the uncertainty in position of the tennis ball Δx must be very, very small in order that the two terms multiplied together give the very small number 6.6×10^{-34} divided by 4π.

It turns out that Δx is so small the uncertainty in position goes unnoticed and the position of the ball can be very accurately seen and measured. Only when we are referring to very small objects on the atomic scale, when the values of Δx and Δp are similar to or smaller than Planck's constant, do the quantum effects become noticeable. This is the beauty of the quantum mechanical theory. It applies to everything whether we are talking atoms or tennis balls. The old classical theory was only useful for tennis balls.

Another interpretation is to consider that a particle does not have separate values for position and momentum but instead has a combination of the two properties, neither of which can be extricated from the combination. So, a distinct value for each can never be derived. We have chosen to use separate concepts for position, momentum and the idea of a particle when these do not really exist in the quantum world. They have been derived from our macroscopic view of the world and we have tried to force them on the quantum world.

Implicit in the uncertainty principle is our inability to predict the future because, if we cannot measure position or momentum precisely, then we do not have any way of predicting what exactly a particle will do in the next instant. It is possible in quantum mechanics and not contradictory to it, however, to work back from current values of position and momentum to determine precisely past values of these quantities. In other words, quantum mechanics reflects our own human view of a defined past but an unknown future.

A future bound up with quantum uncertainty which translates into certainty on a macroscopic scale. We know for certain where a stream of electrons will mostly go when a beam of these particles is controlled in an electromagnetic field but we cannot say for certain that all electrons will go in the same direction. The overwhelming majority will and that is what matters. There is a very small probability that some will not follow but this does not affect the outcome in the macroscopic world. (29)

It is worth noting that if the product $\Delta x \Delta p$ in the uncertainty principle were to be zero rather than $h / 4\pi$, then our classical view of particles would be correct. It would be possible to simultaneously measure position and momentum with complete accuracy. It is because Planck's constant appears in the uncertainty equation that we can conclude that the classical view of nature is incorrect. Furthermore, had Planck's constant been very much larger than it is, then the quantum world in which we live would have been much more apparent to us. (33)

So Max Born's young assistant, Heisenberg, the precocious kid still in his early twenties, had made two major contributions to the new theory. Not only did he show through matrix mechanics that a theory of reality could be constructed without the need to underpin it to classical foundations but now he had also developed a principle of uncertainty which applied not only to atoms and electrons but to the entire realm of our existence. It would take the scientific community some time to come to terms with this second idea and some would say that it still bothers them.

Of all the high-profile players who helped develop the modern and full version of quantum mechanics, including Planck, Einstein, Bohr, Born, de Broglie, Dirac, Pauli and Schrödinger, it is Heisenberg who is considered the star performer. His radical matrix mechanics broke with the classical past and his uncertainty principle ultimately became the centrepiece of the theory. Dying in 1976 at the age of 74, Heisenberg was survived by his wife and seven children. (15)

CHAPTER 6

A NEW VIEW OF REALITY

or how can it be like that?

The Copenhagen Interpretation

Heisenberg's uncertainty principle ensures that we can only ever hope to make statistical predictions. We can never really be sure of the position of any particle, more so as that particle approaches the size of atoms, electrons and photons. As well, when an excited electron decides to drop down to a lower atomic orbit or when an individual radioactive atom decides to decay, can never be predicted. It is odd then that we are able to predict extremely accurately how long it might take for, say, several million atoms of a radioactive substance to decay. The reason this can be done with great accuracy is again because on a large scale the quantum effects disappear. It is very easy to calculate precisely how long it will take several million radioactive atoms to decay because on average this number of decaying atoms behave very precisely. It is only when we look at an individual atom in the population that we see some uncertainty. And this goes for the whole quantum world. On the large scale, we can predict with great certainty when an amount of radioactive material will decay or where a large group of electrons under the influence of a field will end up. It is only at the individual particle level that we cannot say with any certainty what exactly will happen. A very small proportion of individual electrons will not follow the rest but that has no consequence on the larger scale trend.

The situation is very much like the tossing of a coin. We know that if we toss a coin one thousand times we will get roughly 500 heads and 500 tails. We would even possibly be prepared to bet money on such an outcome say to within an error of 100. In fact, it would almost be a sure thing that we would get somewhere between 400 and 600 heads out of one thousand throws. But would we be so sure about the tossing of the coin just once? The odds now aren't quite so good. Now we have only a 50% chance of being right. As it is with coins, the quantum mechanical statistical effects disappear as we use a larger population of coins or atoms or as our particle size approaches the macroscopic size of the everyday world. That is why it took so long to discover the strange behaviour of nature. It was necessary to wait until our instruments were powerful enough to look into this microscopic world. At first, scientists did not like and in fact did not believe what they saw but ultimately they came to accept it.

The older generation had trouble coming to terms with the new theory. The development of quantum mechanics had caused some of these older masters a lot of angst. Albert Einstein did not like Heisenberg's revelations at all and in response wrote to Max Born in 1926 saying: *Quantum mechanics is very impressive. But an inner voice tells me that it is not yet the real thing. The theory produces a good deal but hardly brings us closer to the secret of the Old One. I am at all events convinced that He does not play dice.* (16)

Einstein however could not find fault with the theory which he in fact had taken part in creating. For the rest of his life, he was uneasy with the statistical or probabilistic character of nature upon which the theory insisted. Surely there must be some underlying explanation not yet discovered which would explain everything. Search as Einstein did, none was ever found.

Erwin Schrödinger had similar reservations. With the publication of his wave mechanics approach to the quantum theory, Schrödinger thought he had eliminated the need for quantum steps between the electron energy levels in the atom. He was hoping his theory would show a continuous and smooth increase in possible energy levels for the electron. Schrödinger thought his mathematics could explain what was happening during the quantum jump from one level to another. Working with Niels Bohr in Copenhagen shortly after publication, Schrödinger extended his theory and ultimately showed to his own disappointment that his theory did not preclude the need for quantum jumps between energy levels at all. In frustration, he is reported to have said something like: *Had I known we would not eliminate this damned quantum jumping, I would never have become involved in this business in the first place.* (17)

Interestingly, as the years progressed and they too grew older, both Paul Dirac in 1939 and Louis de Broglie in 1956, expressed their concerns about quantum mechanics and suggested the theory may not yet be totally complete. Even so, the theory today stands much as it was proposed way back in the 1920s. Much has been added but nothing substantial has been changed.

At Göttingen, Born was interpreting the findings as a new view of reality, one that could no longer be considered separate from our own perception of reality. A humanly derived universe? Humans at the centre of the universe? This seemed to be a reversion to very old beliefs but the evidence seemed to be suggesting something like this. Light would behave sometimes as a wave and at other times like a particle and whichever form light chose to be like depended on what application we had chosen for it. If we looked at the photoelectric effect, light was seen to consist of well-behaved particles. If we looked at refraction, light behaved like waves. We seemed to have some control over the form of light simply by observing it. When looking at light, it could be a particle or a wave depending on how we looked at it. When we were not looking, its state was indeterminable.

In 1927, Bohr put all the confusing and often philosophical implications of the theory together in his interpretation of quantum mechanics which eventually became known as the Copenhagen Interpretation since Copenhagen was where Bohr was based. Bohr's statement was driven by a need to explain what quantum mechanics ultimately said about nature. It was a new view of the world.

At the crux of the issue was the process of interaction. By probing the system we are likely to destroy its quantum mechanical behaviour. It is only when we are not observing it that an atom behaves quantum mechanically. Making an observation forces nature's hand to make a decision and adopt one of several possible states available to that system. By state is meant the status of a system—what position or momentum an electron may have at any one time for instance. In going from one state to another, while we are not observing it, we cannot say anything about that system. In fact, we do not know anything about the system. For a single electron going from one state to another, there is no guarantee that the electron in each state is even the same electron!

War Intervenes

As the 1930s progressed, the world was drawn closer and closer to a second world war and, with German researchers dominating quantum research through the 1920s, it would be inevitable that another war in Europe would cut a swathe across the worldwide scientific populations involved in that research.

Werner Heisenberg was being attacked on two sides but not necessarily simultaneously. With the Nazi rise to power in Germany in the 1930s and their persecution of the Jews (of which Einstein was one), many Jewish scientists, including Einstein, were forced to leave Germany and their research ridiculed. It remained for many non-Jewish German scientists, such as Heisenberg, to try to defend their research work. As such, Heisenberg was dubbed a white Jew and subsequently his personal safety became an issue. Then, again, at the end of World War 2, his conduct during the war was questioned by the Allies, mainly because he chose to stay in Germany during the war and accept the post as head of the German research effort into making a nuclear bomb. Many non-Jewish scientists chose to emigrate.

It is well documented that with the Nazi Party entrenched in government in Germany during the 1930s, many valuable research opportunities were relegated to the scrapheap because of the anti-Jewish policies of the Third Reich under Adolf Hitler (1889 – 1945). Research into quantum physics was just one area of research to suffer, with many Jewish German scientists emigrating to Great Britain or the United States of America to escape persecution on religious grounds. Observing this exodus of scientific skills from Germany, Max Planck visited Hitler to suggest that, from a merely practical point of view, the policy of removing Jewish scientists from German institutions did not help the German cause and their subsequent acceptance by overseas research centres in other countries would actually lead to Germany losing the world leadership in many scientific fields. Hitler countered that in relation to the Nazi policy on Jews, there could be no exception. They must be removed. He then flew into one of his characteristic rages and Planck was obliged to leave. Maybe this was a sign that Hitler had no answer to his self-made dilemma.

It is surprising then that Germany, under the Nazi leadership through the 1930s and early 1940s, had pre-empted the rest of the world by decades in relation to research and policies on such issues as alcohol and tobacco consumption as well as banning the use of asbestos for insulation in building materials. Tobacco, asbestos and alcohol were deemed to be carcinogenic, with restrictions put in place to control or eliminate their use. This was decades ahead of similar campaigns in Great Britain, the United States and many other countries. Non-smoking train carriages were provided and smoking was banned in public areas, along with cigarette advertising. It was even possible to be arrested for causing a car accident while smoking, something that might happen today with the use of mobile phones. Economic issues also played a part in these health policies. It was pointed out that the production of meat required a far greater input of resources and energy than simply growing grain for consumption. The fact that Hitler did not drink alcohol, did not smoke and was a vegetarian, gave weight to these policies under his dictatorship. These forward-thinking ideas did not however make up for the irrational policies the regime would make during its short 12 year reign. (30)

In 1939, war was declared against Germany by Britain and France. Initially, these two countries were no match against Germany, a country that had been fortifying itself with modern war machinery and tactics for years but, once Germany's blitzkrieg, or lightning war, failed to win a quick victory for Germany as anticipated, World War 2 morphed into a war based on rates.

Firstly, there was the rate at which destroyed equipment, such as tanks and aircraft, could be re-supplied. If you did not have the capacity to rebuild at a rate faster than your opponent then you would eventually lose. Secondly, there was the rate at which new designs and secret weapons could be developed and brought online to assist the war effort. Many areas of scientific research were called upon to assist in this development and in this arena Germany was ultimately the loser.

Before the war, Germany was a leader in atomic research, jet engine design and rocket technology and all of these could potentially be extremely useful to the war effort. From the outbreak of hostilities in 1939, the Nazi regime in Germany failed to capitalise on this fact. As well, there was the consideration that the rate of development of new miracle weapons based on these technologies would not be quick enough for them to be deployed before the war ended. So nuclear research in Germany was not given priority and this was in stark contrast to how it was handled by the Allies.

Upon the realization of the potential of nuclear energy for the production of large bombs, the United States, in 1942, set up the Manhattan Project, specifically to develop the nuclear bomb. It was given the name Manhattan Project because initial planning talks regarding its aims were held in New York. Ultimately, there were more than 100,000 personnel working on the project compared to a few thousand in Germany. Generally, Hitler could not grasp the potential of these new weapons and saw the war more in terms of the traditional weapons employed in World War 1, albeit using modern improved versions. Hence, like it or not, Heisenberg found himself in charge of a research project with limited funds and a certain lack of urgency to do anything about the potential of nuclear energy applications during the war. On the other hand, jet engine development and rocket science were given top priority by Hitler but weapons using these new technologies suffered from either low production rates or poor choice of application, both of which limited their effectiveness on the battlefield.

In 1941, at the height of the war, Heisenberg travelled to Nazi-occupied Denmark to talk with Niels Bohr about recent developments in atomic research. The conversations they had have been thoroughly analysed but much doubt still remains over what was said and the conclusions drawn by both men. Was Heisenberg attempting to recruit Bohr into the German nuclear research program? Was he announcing Germany's position and intent to build a nuclear bomb and therefore purposely giving impetus for the allied program to proceed at full speed to do the same and hopefully beat Germany in the atomic race?

Years later, after the United States had successfully deployed two nuclear bombs over Japan, Bohr was still troubled over whether his interpretation of what Heisenberg was saying about Germany's intent to make the bomb gave impetus to the Allied bomb research culminating in the subsequent destruction of Hiroshima and Nagasaki and their large human populations. It was only towards the end of the war that the Allies realized that Germany was well behind in their own nuclear bomb program and there was no race at all but, by then, the construction of the nuclear bomb at Los Alamos in the US had reached such momentum that the program was virtually unstoppable.

Much discussion centres on whether the German scientists had the knowledge and the motivation to build a nuclear bomb during the war years and it seems that they did not. The program to build a bomb was never given the same priority as was, say, given to the project to build the V1 and V2 rockets under Wernher von Braun (1912 – 1977).

After the war, many of the German scientists involved in nuclear research were rounded up and detained in England for some time at Farm Hall near Cambridge where their conversations were secretly recorded and analysed. A fair assessment of Heisenberg seems to be that he was not a Nazi. Instead, he was a staunch nationalist and supporter of anything German. It seems that he may well have had little interest in politics, burying his thoughts in his love of physics. Having few skills or aptitude for experimental research, it was the theoretical that interested him most. Surprisingly, his mathematics was often sloppy.

It appears Heisenberg never met Hitler although, in 1937, he did have a brush with Heinrich Himmler (1900 – 1945), the Reichsführer-SS who was head of the ruthless security forces of the Nazi Regime. Heisenberg had supported Einstein's theory of relativity. Einstein was Jewish and so his scientific research was unacceptable to Hitler's Third Reich. Ultimately, and after a thorough investigation, Heisenberg was cleared, Himmler stating that *Heisenberg was a harmless, apolitical academic for whom theoretical physics is merely the working hypothesis with which the experiment inquires of nature . . . a man of decent character.*

Heisenberg was convinced the war would not last long and even discussed this view with Albert Speer, Hitler's Minister of Armaments. Albert Speer (1905 – 1981) had come to prominence in the Nazi regime after he was appointed Minister of Armaments under Hitler in 1942. Speer was 37 years old. He had been Hitler's personal architect put in charge of designing a new built environment for Germany. He clearly had great organizing skills and quickly assessed what Heisenberg was telling him.

In Heisenberg's opinion, America had now taken over the lead in nuclear research—a discipline dominated by Germany before the war and closely related to atomic and quantum physics but dealing specifically with the nucleus of the atom rather than the electrons which surround it. Due to the significance of this research in potentially producing a bomb of tremendous size, Speer offered Heisenberg virtually any resources he needed and was dismayed when Heisenberg requested so little. It would take three to four years to develop a bomb, he said, by which time the war would be over. Heisenberg switched his research to concentrate on building a reactor, a device for producing heat which could be used to produce steam to drive machinery via a steam turbine. Such a reactor is used today in power stations and nuclear submarines.

A radioactive substance is one where the nucleus of its atoms spontaneously breaks up so creating two quite different lighter elements as well as emitting several neutrons and possibly other sub-atomic particles. This nuclear reaction is in stark contrast to the everyday chemical reaction where only the electrons surrounding the nucleus come into play. The nucleus of the atom is left untouched. Heat energy can be produced in a chemical reaction, such as in the burning of a hydrocarbon, but nothing like the heat produced in a nuclear reaction.

In nuclear reactions it is the composition of the nucleus which is of interest. Made up of protons and neutrons, it is these nucleons which define an element and its variations or isotopes. All isotopes of uranium, for example, have 92 protons in their nucleus since 92 protons defines the element as uranium. If the number of protons was different from 92, it would be a different element. Importantly, it is the number of protons that defines an element. Isotopes or variants of uranium may have slightly different numbers of neutrons. So there is Uranium-238 with 146 neutrons and Uranium-235 with 143 neutrons, both with 92 protons. The total number of nucleons (protons plus neutrons) is written after the symbol for the element—for example Uranium-238 or simply U-238.

Uranium-238 is the most stable isotope of uranium while U-235 (with 3 fewer neutrons) is more unstable and is the form used for making nuclear bombs. The nuclei of the atoms of a radioactive material, from time to time, will spontaneously break up and emit several neutrons. That is what makes them radioactive. They can also be encouraged to break up by firing neutrons at their nucleus. In the case of U-235, bombarding this isotope with slow neutrons has this effect while with U-238 very fast high energy neutrons need to be used. In either case, the uranium atom breaks up into two different elements plus several free neutrons. If one of the neutrons emitted when, say a U-235 atom decays, strikes another U-235 nucleus, the nucleus will split and emit two further neutrons. In turn, these two neutrons will strike other U-235 nuclei which will then release further neutrons, and so on, producing a chain reaction.

In a flash, all the uranium atoms of the sample of uranium will split. In a power station, this chain reaction is controlled by surrounding the uranium with control rods which absorb neutrons and so slow the chain reaction. In a bomb, the chain reaction is uncontrolled. As well, the uranium fuel is surrounded by a moderator such as water, graphite or heavy water which slows the neutrons, so ensuring they split another U-235 nucleus, a process that requires slow neutrons.

The proportion of U-235 in naturally occurring uranium is less than 1%, the rest being the less-volatile U-238, so much difficult refining is necessary to produce bomb-grade U-235 uranium. Plutonium is also used as an alternative to uranium. As well, it is important to realize the mixture of uranium and the amount used in a reactor is such that, in the advent of an accident, the reactor may overheat, melt down and release much radioactive material but it cannot explode like a nuclear bomb.

For a spontaneous chain reaction in a bomb to occur, there needs to be a certain amount of uranium to ensure the free neutrons actually have a chance of striking another uranium nucleus and not simply escape from the material altogether. It turns out that the critical mass or amount of uranium needed is of the order of only a few kilograms. A nuclear bomb is detonated by quickly bringing together separated pieces of the critical amount. In the bombs dropped on Japan in

World War 2 this was achieved in two ways. Using conventional explosives, the gun method fired one piece of uranium into another so creating a critical mass. The implosion method used conventional explosives placed around the outside of a spherical sample of low density uranium. Upon ignition of the conventional explosive, the ball of low density uranium is compressed, creating a sample of high density uranium greater than the critical mass.

What causes the nuclear bomb to be so destructive? The heat energy of the bomb comes from the fact that some mass is lost during the fission or breaking up of the uranium nucleus into two new nuclei. The mass of the initial radioactive nuclei is greater than the sum of the masses of the two resulting nuclei and any free neutrons emitted. This lost mass appears as heat energy and its value calculated using Albert Einstein's famous equation $E = mc^2$ which simply means a very small mass is equivalent to a very large amount of energy. This is because c, the speed of light, is a very large number and even larger when it is squared. What this equation shows is that a small loss of mass translates into a massive release of energy.

An issue that bothered both the Americans and Germans who were involved in their respective nuclear bomb programs was the possibility of an uncontrolled chain reaction. Once the bomb ignited, would it stop when the uranium was depleted or would the chain reaction take off through the atmosphere of the entire planet? Apparently, Hitler was concerned the earth would become a glowing star. (31)

After World War 2, Heisenberg was appointed director of the Max Planck Institute in Göttingen and remained there until his death in 1971. He was appointed a Fellow of the Royal Society in London and was mostly accepted by the worldwide scientific community. However, his relationship with Niels Bohr was never truly repaired. Bohr died in Copenhagen in 1962. Max Planck, who had begun the quantum revolution, was a tired old man by war's end. His elder son Karl had been killed in World War 1 and his twin daughters died giving birth. Both grand-daughters survived. His second son, Erwin, was executed by the Nazi regime in early 1945 for supposed complicity in the attempt to assassinate Hitler in 1944. Planck had a third son, Hermann, to his second wife in 1911. Allied bombing destroyed his home in Berlin and, in the closing days of World War 2, he was rescued by the Americans after hiding out in forests from

Richard Feynman in Geneva, July 1958.

(AIP Emilio Segrè Visual Archives, Segrè Collection)

possible capture by the Russians. Planck died two years later in Göttingen in 1947.

Anti-matter and Time Travel

The war years had faded into history. The groundwork in the quantum theory, which now encompassed a comprehensive description of the entire atom and hinted at further revelations about reality itself, had been completed but there was no stopping the theory's continuing development into an even more comprehensive description of nature. So, when the phone rang at four in the morning on October 21, 1965, the American physicist Richard Feynman (1918 – 1988), working at the California Institute of Technology (Caltech), knew his time had come. He at last had joined the long ranks of quantum pioneers to receive the Nobel Prize—Planck (1918), Einstein (1921), Bohr (1922), de Broglie (1929), Heisenberg (1932), Schrödinger (1933), Dirac (1933), Pauli (1945) and Born (1954).

At home in California, Feynman briefly told his wife the exciting news (she at first did not believe him) before trying to get back to sleep. He knew that would be impossible. The journalists just kept ringing. Upon receiving the Nobel Prize, one's life is irreversibly and forever changed. As usual, the newspaper reporters had got wind of the announcement five hours before the official telex message came through from Stockholm at a more civilised breakfast time. Feynman along with two other researchers, Julian Schwinger and Shin'ichiro Tomonaga, the message read, had been awarded the Nobel Prize for physics:

Professor Richard Feynman Physics Dept
California Institute of Technology Pasadena (Calif)

Royal Academs of Sciences today awarded you and Tomonaga and Schwinger jointly the 1965 Nobel Prize for physics for your fundamental work in quantum electrodynamics with deep ploughing consequences for the physics of elementary particles. Prize money each one third. Our warm congratulations. Letter will follow.

Throughout the early and still dark hours of morning a steady stream of news crews and photographers arrived at the house. The phone just kept on ringing so eventually it was left off the hook. By this time Feynman had become quite adroit at responding to questions, many of them trite. As usual the newspapers, radio and television stations only wanted a short and mostly meaningless grab of a few seconds duration to fill a news spot. Feynman realized it was useless to try to answer the questions seriously. The reporters' ears were closed to that. They were interested only in personalities. Tired of answering questions like briefly tell us what you did, Feynman finally used the retort actually suggested to him earlier by another reporter—*Listen buddy, if I could tell you in a minute what I did, it wouldn't be worth the Nobel Prize!* (18)

Feynman's work centred on one of the initial aspects of the quantum theory first investigated by Planck way back in 1900—namely the emission of a photon from an electron, hence the name of this field quantum electrodynamics or QED for short. During World War 2, Feynman had worked on the Manhattan Project at Los Alamos in New Mexico where the nuclear bomb was developed. In the late 1940s, he completely reworked the quantum electrodynamical part of quantum mechanics to come up with a much better version of this section of the overall theory. It took nearly twenty years for the value of his work to be recognized and for Feynman and his two associates to be awarded the prize.

Feynman's approach was to avoid as much mathematics as possible and, instead, present interactions between particles by way of graphs such as that shown in Figure 23. To simplify things, these graphs utilize a one-dimensional universe (instead of three dimensions) where particles can move back and forth along a single line depicted by the horizontal axis. Time is shown along the vertical axis. Figure 23 shows a simple collision between two particles drawn on a Feynman diagram or as they are more generally known now, a space-time diagram. Particle 1 comes in from the left and Particle 2 from the right. They collide at A and then rebound.

There are some important points described in this deceptively simple-looking graph. Firstly, remember that for simplicity we are using a one-dimensional space rather than a three-dimensional space so the collision described is actually a head-on collision. Both particles rebound along the same path along the horizontal axis that they entered along although this doesn't appear to be the case. Remember the vertical axis is time and not another space dimension. It is used to spread out the collision event so the incoming and outgoing paths of the particles do not overlap on the graph. By doing this, some further details of the event can be gleaned. Particle 2 rebounds at a faster speed than Particle 1 because its rebounding path is closer to the horizontal. This is so because, in a shorter space of time indicated by the vertical axis, Particle 2 travels further away from the collision point A. The rebound path for Particle 1 is nearly vertical indicating that it is rebounding at a very slow speed. If the path was completely vertical, the particle would not be moving at all—its value on the space axis would remain constant for all values of time (the definition of

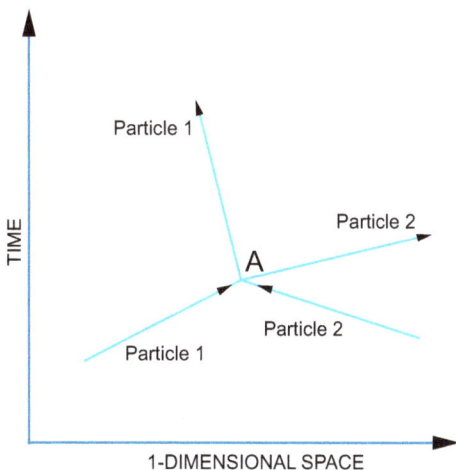

FIGURE 23: A Feynman or space-time diagram showing the collision of two particles at A and their subsequent rebound back along their 1-dimensional space world, albeit at different speeds shown by the different gradients of the paths.

a stationary particle). A horizontal line would indicate infinite speed since the particle would then be changing position with no lapse in time—an impossibility since nothing can travel infinitely fast, the speed of light being the fastest possible speed for any particle. Einstein showed this in his theory of relativity published in 1905. So, all lines on a Feynman graph depicting the paths of particles through time must be inclined to the horizontal. They are called world lines since they describe the history of a particle through time.

One way of viewing the collision described above is to cut a window slit in a piece of paper and slide it up the graph. Through the slit you will see the particles coming together and moving away exactly as described in the preceding paragraph. The graph is like a series of frames on a movie film although in the case of the graph the transition between frames is smooth. Sliding the paper slit up the graph is like running the movie projector. The still images become animated.

Figure 24 shows a real-life interaction where an electron comes in from the left, emits a photon of light and recoils back towards the

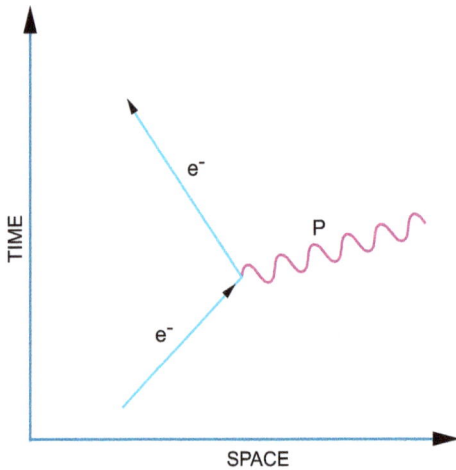

FIGURE 24: A Feynman or space-time diagram showing a real-life situation where an electron comes in from the left, emits a photon of light and recoils back to the left. In a Feynman diagram, an event's activity is read beginning from the bottom of the graph and moving up the diagram with time.

left. Feynman's QED theory was full of graphs like this describing all sorts of sub-atomic particle interactions. Using them he was able to develop a logic which helped sort out the mathematics for his theory of quantum electrodynamics.

Earlier, however, and before America's involvement in World War 2, Feynman worked as a young assistant to physicist John Wheeler (1911 – 2008) at Princeton University where he was studying for his PhD degree. This was the early 1940s. Specifically, Wheeler and Feynman's work involved the possibility of anti-matter and backward-in-time events. Around 1930, Paul Dirac had suggested the idea of a particle he called the positron—a particle the same size and weight as an electron but with a positive rather than an identical but negative charge. The existence of such a particle was predicted by his theoretical calculations and within a few years of his suggestion, experimental scientists detected such particles. Eventually it was accepted that all the fundamental particles of the atom such as protons

116

Informal portrait of John Archibald Wheeler at Copenhagen, Denmark.

(AIP Emilio Segrè Visual Archives, Wheeler Collection)

and neutrons also had anti-particle equivalents. The whole collection became known as anti-matter.

Positrons can be produced, together with their electron counterparts, when high energy light rays known as gamma rays strike matter. In quantum terminology, one photon of light creates one electron-positron pair of particles. Positrons are short-lived and this is why we do not normally see them. If they strike another electron they quickly combine with it to reverse the process and so produce a short burst of light, a photon, in their place.

Using a space-time diagram, this process is depicted in Figure 25a. This describes Photon 1 moving through time, coming up from the base of the graph and, at some point A, creating an electron with negative charge, e⁻, and a positron with positive charge, e⁺. The electron, e⁻, goes off to the left into the greater universe but the positron, e⁺, heading to the right encounters another electron, e⁻, after a short time at point B. The positron and electron combine to produce

117

Photon 2. In standard quantum theory there is nothing very exciting about this but the often bizarre Feynman suggested a radical plan. He redrew the diagram, replacing the positron with an electron that is effectively travelling backwards in time. In doing this, he was able to make some profound deductions, but first let us look at the diagram (Figure 25b) as he drew it.

Feynman suggested the process described in Figure 25a could be equally described as shown in Figure 25b. Rather than a photon coming up from the bottom of the graph and, at point A, annihilating to produce an electron-positron pair, the point A could be described as an electron simply bouncing off Photon 1. This photon is annihilated but a second (Photon 2) appears at a slightly different place B where the same electron has just been deflected by it. The only difference between the two cases is that the positron is replaced by an electron travelling in the opposite direction. This direction is backward in time since its arrow is pointing towards the base of the diagram.

Feynman argued that a positron could be considered as simply an electron travelling backward in time. But there is more. Figure 25b describes the same situation as that of Figure 25a but uses only one particle, an electron rather than two particles. In using only one particle to describe the events it is necessary for that particle to travel back in time so that it can be at certain places at the required times. And more still. Feynman argued if one particle has the ability to take on the work of two particles simply by having the capability of moving backward in time, then surely this same particle could take on the work of even more particles by continually darting backwards and forwards through time. Taken to the extreme, this one electron could do the work of all electrons in the universe. He argued that there is really a need for only one electron in the entire universe and in fact all the electrons we see are really the same electron and this explains the reason why all electrons are identical. Carrying the argument to its logical conclusion, the entire universe is made up of a single proton, neutron and electron since these three fundamental particles are the building blocks of any atom in any substance.

Further, if all the particles of the universe can be described in this way, then Figure 25b is how the universe could be represented on a space-time diagram. This shows all the particles and light rays of the universe interacting through time. If the paths or world-lines of all

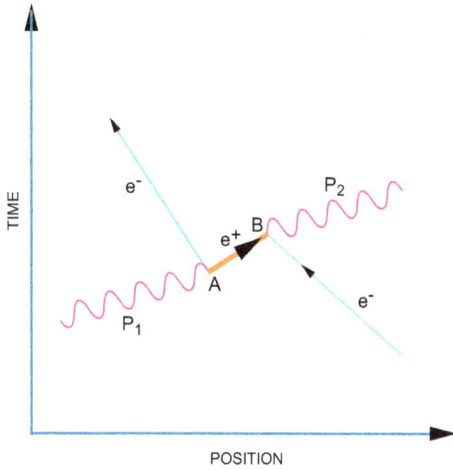

FIGURE 25a: A space-time diagram showing an electron/photon interaction with the creation of a positron. Photon 1, moving upwards from the bottom left of the graph, creates an electron/positron pair at A. The electron moves away to the top left and, at B, the positron merges with another electron coming from the bottom right to form Photon 2.

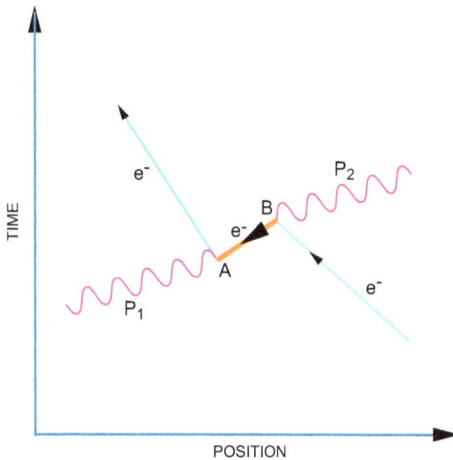

FIGURE 25b: The same interaction as shown in Figure 25a but with an electron travelling backwards in time replacing the positron. This interaction requires just one electron rather than an electron/positron pair but requires that electron to travel backwards in time—the path between B and A which slopes down.

119

particles are fixed in this way in space-time then the flowing of time might be taken to be a human illusion. Maybe it is only us that sees time flowing. Again, using a slit cut in a piece of paper and moving this slit slowly up the diagram of Figure 25b we see a hive of activity amongst the particles and light rays.

It is only conjecture that positrons are electrons moving backwards through time but such an idea does not conflict with any laws of physics. No particle has yet been detected violating our sense of the forward movement in time, yet to propose such ideas is extremely useful in scientific thought. It is often these outrageous thoughts that are the trigger for great discoveries and it was some of these supposedly outrageous thoughts which helped Feynman a few years later to come up with a completely new way of looking at quantum electrodynamics. The Nobel Prize honours such original and useful thought and in 1965 it was Feynman's turn.

End of an Era

Let's recap. By the 1930s, the basic structure of the atom and the theory behind it, quantum mechanics, was well and truly established. The pioneers of the 1920s, many of them German, were now famous and middle-aged but, as has already been mentioned, a dark cloud was forming over Germany and indeed over all of Europe with the rise of the National Socialist Party or Nazis. Hitler gained power in 1933 and accordingly free speech in all disciplines including science was stifled. The atmosphere at Göttingen changed overnight. An atmosphere which encouraged free thought and questioning of the status quo was now spurned by those who thought they knew the way forward. Quietly, intellectuals including many physicists began to leave Germany.

Einstein got out early. Others were to follow. Many physicists emigrated to the United States where they became the obvious choice to work on the Manhattan Project—the codename for the American research effort to develop the nuclear bomb at Los Alamos in the New Mexico deserts. To the new and younger generation of theoretical physicists, the war came as an interruption to their studies and Richard Feynman fell into this latter category.

*Richard Feynman's
badge photo from
Los Alamos National
Laboratory.*

*(Los Alamos National
Laboratory, courtesy
AIP Emilio Segrè
Visual Archives)*

In the early 1940s, he was working on his PhD project involving the nature of time when he was visited one day by an official whose job it was to recruit bright researchers for the Manhattan Project. For security reasons, the official was not supposed to tell anyone what the job entailed but he knew that if he wanted to enlist Feynman he would have to divulge the secret. He considered this wouldn't be a problem because once Feynman knew what the project was, he would certainly jump at the opportunity to be involved.

He was wrong. Feynman said he wasn't interested. The official, knowing Feynman better than Feynman, accepted the refusal and left the room, quietly mentioning there was a meeting at three and he would see Feynman there. Needless to say, Feynman, within a few hours had changed his mind and had decided to turn up. By four o'clock that same day he was working on details of the nuclear bomb project. (20)

The 1920s, 1930s and 1940s were the halcyon days of the quantum theory, when major intellectual leaps were made that resulted in the fast development of the theory. But just as often as useful and practical discoveries were uncovered, challenging and unsolvable puzzles seemed to arise. Chemical and nuclear reactions had been explained, as was the structure of the periodic table of elements yet the quantum theory also implied some hard to accept revelations about the makeup of our universe and these included the EPR Experiment, Schrödinger's Cat and the Many Worlds Interpretation.

Double Slits Again

As we have seen, the Schrödinger equation gives a value for the state function Ψ of a particle at any instance. The equation describes the state function as this function evolves through time but the state function does not have any real or physical meaning. In fact, it is quite puzzling as to what it really represents. Squaring the state function at any time does however give us something more concrete to work with. The square of the state function Ψ^2 produces a probability of finding the particle at a specific place.

There are two issues here. The state function Ψ can be considered a description of a particle which is smeared or spread throughout space. Then, when we need to make a measurement to determine the particle's position, we apply Ψ^2 to give us a probability distribution for finding the particle in specific places. Making a measurement and so obtaining a probability from Ψ^2 has become known as *collapsing the state function*. This so-called collapsing provides us with some measure of a concrete description of the particle's whereabouts at the moment we make the measurement. The moment the measurement is over, the state function for the particle again begins to smear or spread out. In other words, every particle is spread out over a large region until the moment we wish to measure its position. At that moment, we somehow collapse the state function to give us a precise probability function for finding a well-defined (as distinct from spread out) particle at a particular place.

Recall the double slit experiment where we fired just one photon or one electron (it doesn't matter which small particle is used) at a time towards the double slits and allowed a distribution to build up on the screen. The state of each particle on emerging from these double slits is now described by a state function with two peaks in its amplitude, namely $\Psi = \Psi_1 + \Psi_2$. This combined state function for the emerging particle Ψ describes the particle as being spread out everywhere but, in particular, more closely associated with the two slits. In effect it describes a single particle being in two places at once. It does not describe a photon or electron which has split in half with each portion going through either slit because photons and electrons have been observed to never split in this way.

Recall also that if we make a measurement at one of the slits to determine which slit the particle passes through, the interference pattern on the screen disappears and in fact if this measurement is made, the photon or electron does actually pass through only one of the slits. By making the measurement we are again collapsing the state function and through Ψ^2 obtaining a probability distribution and ultimately determining where the particle is. At that moment, a new state function is created and the particle again begins to spread out from that slit. Now however no interference pattern will be observed because the particle did only go through one slit and for interference to occur both slits need to somehow be involved.

The non-interference pattern that we obtain when we observe which slit the particle went through is the same pattern we obtain when one of the slits is closed—just a diffuse glow on the screen adjacent to the open slit. Opening the second closed slit again creates the interference pattern. Closing a slit or placing an instrument at the slits to indicate which slit the particle went through again eliminates the interference pattern.

It could be argued that the act of placing a sensor at the slits to measure the passing of a particle actually interferes with the passing particle in such a way that the experiment is now essentially altered. To avoid this, it is possible to build more and more delicate sensors so that the disturbance is minimal but ultimately a limit of refinement is reached where it is impossible to go any further. This is because of the Heisenberg uncertainty principle which proposes, as we have seen, a limit in the knowledge about position and momentum of a particle and hence a limit to the refinement of any measurement we could possibly imagine trying to make.

In a sense, the difference in probability distributions for the two situations (interference and non-interference) is the difference between $(\Psi_1 + \Psi_2)^2$ and $\Psi_1{}^2 + \Psi_2{}^2$.(19) When an event has several different but possible ways to happen then the state function, Ψ , or the probability amplitude which was used in a similar way previously, is the sum of the individual functions for each alternative, namely $\Psi_1 + \Psi_2 \ldots$ etc. As was also mentioned previously, the probability for the event becomes the square of this, namely $(\Psi_1 + \Psi_2)^2$ and the interference pattern is observed. On the other hand, if the experiment is conducted in a manner that allows determination of which alternative was actually taken, then the probability of the event becomes instead the sum of the individual probabilities for each alternative, namely $\Psi_1^2 + \Psi_2^2$ .. etc. and the interference pattern disappears.

Recall before, when probability theory was used to analyse the behaviour of the double slit experiment, when complex numbers are involved (which they often are in quantum mathematics) the expression $(\Psi_1 + \Psi_2)^2$, when multiplied out contains a term containing the mathematical function $cos\ \theta$. It is this term which generates the interference effect because the value of $cos\ \theta$ oscillates between 1 and -1 as θ is increased, so producing a fluctuating cosine wave. Check this out on a calculator. The case where each alternative is simply squared, namely $\Psi_1^2 + \Psi_2^2$, contains no such $cos\ \theta$ term and hence no interference pattern.

So, what is going on? Nobody really knows. It seems that the particles not only know when either one or both of the slits is open but also know when we are looking at them. The particles then adjust their behaviour accordingly. Essentially, what the preceding experiment is saying is that when you do not observe nature it does one thing but the moment you look at nature it behaves differently. We are back to the worrying situation of human intervention. Even if you do probe particles to see which slit they passed through, there is still the more fundamental consideration of trying to predict which slit they will pass through in the first place. No one has worked that one out. Perform the experiment over and over and the results are always different.

Throughout the ages, science has always been proud of its predictability. Set up an experiment and it will perform repeatedly like clockwork. Our whole physical world appears to run like clockwork but now when we look at the finer details on the level of the basic building blocks (the atoms), we discover unpredictability. A world run by statistics and probability. But don't panic. The marvel of it all is that this microscopic quantum world of roulette translates into a macroscopic world of total predictability where nothing really happens by chance. Chance averages out to precise and determinable outcomes. In other words, the microscopic probabilistic world forms the building blocks of our macroscopic deterministic world with which we are so familiar. Quantum theory describes both the microscopic and macroscopic worlds whereas the old classical theories of physics only describe the macroscopic world. The situation is similar to that where the old Newtonian description of nature could only describe a world with slow speeds whereas the relativity theory developed by Einstein described nature at any speed.

The EPR Experiment

Even though the quantum theory was deemed highly successful and useful in its description of the workings of the atom, the behaviour of particles in the double slit experiment still led to some worrying implications—implications which seem to flout the rules of classical physics and tear apart our well-constructed view of the world. It was these issues which worried Einstein to the day he died.

The implications of quantum mechanics are far reaching and are integral to our view of the world and indeed the universe, even though quantum mechanics appears to be only of use when discussing the micro-world of the atom or in esoteric experiments like the passing of photons or electrons through double slits. The point of using atoms and double slits is to illustrate aspects of the quantum theory. These double slit examples are merely vehicles to illustrate the behaviour of the universe at large according to the quantum theory. The double slit experiment is one of the easiest demonstrations to set up and it is assumed that if particles and waves behave strangely in this very tangible arrangement then they must also behave similarly in the wider universe. The implications of this are quite disturbing.

For instance, the double slit experiment seems to imply that information can travel faster than light. No wonder Einstein was concerned. That the speed of light is the maximum possible speed is fundamental to Einstein's theory of relativity and now here was an experiment that seemed to contradict this.

When a device is placed at one slit to determine whether a particle went through it or not, it could be immediately determined what event occurred at the other slit. In the experiment where one particle at a time was fired, if a particle went through slit 1, then we know the particle did not go through slit 2. More importantly, it is the fact that a particle went through slit 1 which instantly determines no particle at slit 2. The moment we detect an event at slit 1, slit 2 knows what to do so, in effect, information travelled between the two slits instantaneously so violating Einstein's speed limit.

The argument against this paradox is based on the cause and effect approach. An electron at slit 1 did not directly cause a lack of an electron at slit 2 so it could be argued that information did not travel between the two slits. The possibility of a particle appearing at either slit is bound up in the state function of the particle—both cases co-exist until we make the observation and so collapse the state function. This is Niels Bohr's Copenhagen Interpretation.

The situation is a bit like this more familiar scenario. I am undecided about buying an item—say a car on display in a showroom. That item is in a state of being bought or not being bought. The moment I decide, say, not to buy (that is, my state is now determined), the car's state is instantly determined as well, no matter how far away from me is the showroom. At least from my position, I know the state of the item, in this case the car, has changed.

The concern about this paradox became so strong that three researchers proposed a thought experiment—an experiment perhaps unable to be performed for practical reasons but one which could be imagined to test a hypothesis. This particular thought experiment, devised in 1935, became known as the EPR experiment after its three proponents—Albert Einstein, Boris Podolsky (1896 – 1966) and Nathan Rosen (1909 – 1995). It was an attempt to refute Bohr's Copenhagen Interpretation which implied *spooky action at a distance* as Einstein put it. (21)

The experiment involves two particles A and B which interact or collide and then fly apart. At any time, each particle will have its own momentum and position in space. It is quite allowable in quantum physics to know the total momentum of a system exactly but not, as we have seen, make accurate simultaneous measurements of position and momentum of a single particle within that system. That would violate Heisenberg's uncertainty principle, the central foundation of the quantum theory. It was this principle that Einstein and his cohorts

Portrait of Werner Heisenberg in middle age.

(Max Planck Institute, courtesy AIP Emilio Segrè Visual Archive)

attempted to refute. Einstein was very much against the uncertainty principle as well as instantaneous action at a distance because it flouted his discovery that nothing could travel faster than light. The speed of light was finite, not infinite, and an infinite speed would be required for instantaneous transfer of information.

Einstein suggested simultaneously measuring the momentum of particle A and the position of particle B. These two measurements could be precisely made since they referred to different particles, not the same particle, so these measurements did not contravene the uncertainty principle. Knowing the momentum of A implied knowing precisely the momentum of B since we knew the total combined momentum of the system. To obtain the momentum of B, all we had to do was subtract the momentum of A from the total momentum of the two particles. Hence, we now know precisely the position and momentum of particle B in direct contradiction to the uncertainty principle which stated it was not possible to accurately know the momentum and position of a single particle. Onlookers were aghast. Was this the downfall of quantum mechanics after only a ten-year reign?

There was no simple answer and arguments raged for decades and well past the death of Einstein himself in 1955. Einstein was arguing that the quantum theory was incomplete. There must be some underlying hidden variables which removed the need for the uncertainty principle. They just had yet to be discovered. Einstein was not saying quantum theory was wrong, only that it needed additional development. He argued that if by measuring the momentum of particle A we could infer the momentum of particle B then, since we do not interfere with particle B in any way, it must have had that definite value of momentum all the time irrespective of our tampering with Particle A. Particle B must therefore have properties not described by quantum theory that maintain its value of momentum before and after measuring the momentum of Particle A, so it could be argued that the theory of quantum mechanics is not yet complete. In effect, Einstein was saying that whatever we do to Particle A could not have any instantaneous influence on the distant Particle B. The physical influence of measurement was localised to the current particle. In Einstein's words, there was no spooky action at a distance.

Bohr's interpretation was very different. The Copenhagen view was that a particle had no definite position or momentum until a measurement was made. After measurement, a definite value for position or momentum came into existence from an indeterminate fog of possibilities. (22) This was the Copenhagen Interpretation on which all the quantum theory was built. There were no underlying hidden variables which produced definite values of position and momentum. The quantum world was essentially probabilistic and no one could produce a theory using hidden variables that could generate the very real and useful outcomes of the existing quantum model. Bohr countered Einstein by suggesting that measuring the exact position of A so that the exact position of B could be deduced from Particle A meant that the momentum of A could not be accurately measured and therefore neither could the momentum of B be accurately calculated. The conversations between Bohr and Einstein reached a feverish pitch.

A variation of the EPR thought experiment involves the quantum property of spin. If a hypothetical quantum system consisting of a particle with zero spin suddenly divides into two particles, the sum of the spins of the two new particles must add up to be zero. The particles could be, say, electrons. It is quite all right to know the exact spin of a system—in this case, zero. This is the conservation of spin law, similar to the conservation of energy and momentum principles. If one particle spins clockwise then the other must spin counter-clockwise and there is no knowing which way either will spin until a measurement is made. Like many features of the quantum world, spin allocation is random.

Now as both new particles fly apart, quantum mechanically speaking, they both have a superposition of states of spin—either particle could be spinning in either direction. At a later stage we measure the spin of one particle. At this moment the state function for that particle collapses to give a definite value of spin. Through the conservation of spin principle, the state function for the other particle must collapse at precisely the same moment to give a value of spin opposite to the first. There is instantaneous action at a distance no matter how far apart the two particles have travelled even if it is to opposite sides of the universe. This has become known as a non-local influence.

Although this seems a contradiction to the belief that, according to Einstein's relativity theory, it is impossible to transfer information instantaneously over a distance, the spin experiment in no sense makes this type of transfer. The initial spin is allocated randomly upon creation of the two particles and so measurement of the spin of one particle only produces a random value for the other particle. This is not a useful information transfer. What can be inferred from the thought experiment is that once particles have interacted or been simultaneously created, they are forever somehow bound together within a single quantum system no matter how far apart they might stray. Even at opposite ends of the universe they are still somehow one.

This situation is called quantum entanglement and was first proposed by Erwin Schrödinger. It refers to two particles that have interacted in some way, such as a collision, and then moved well apart. Entanglement is an important part of the developing theories of quantum computing and teleportation. Because the particles have interacted, they remain linked so if we measure the spin of one particle, the value is random, but you instantly know what the spin of the other particle will be. The value of the distant particle is not random but because of the initial randomness there is no transfer of useful information. and you cannot in any way use the link between the two particles to send useful information. Einstein's theory of relativity is upheld—neither energy nor information can be transferred at a speed faster than light.

In summary, Einstein's EPR experiment implies any influence on a particle during measurement is confined to the particle or, in other words, is local. Proponents of the Copenhagen view claimed influence to be non-local—a local measurement can affect a distant one instantaneously but with no information transfer. Einstein's view was closely related to our everyday view of reality—that there are things which exist which we would call real even when we are not looking at them and no information or influence can travel between two places faster than the speed of light.

Since its debut in 1935, the proposed EPR thought experiment has been conducted in many different ways, each claiming to be a perfect test of the issue. Ultimately, each of these tests show Einstein to be wrong. There does seem to be spooky action at a distance but with each experiment there seems to always be a minor snag, a way out for the Einstein camp. The case against the EPR experiment is not yet closed but it would be a brave researcher who staked his future on believing in it. The evidence seems to weigh heavily against Einstein and our realistic view of nature. Except for the absence of gravity in the theory and other issues such as dark matter, it seems that quantum mechanics does provide a complete description and there is no need for hidden underlying variables yet to be discovered which can explain away the fuzziness of reality while not being observed. This view however continues to be challenged. As we shall see, David Bohm in the 1950s, John Bell in the 1960s and Allain Aspect in the

1980s investigated the dilemma further with all outcomes supporting the Copenhagen camp.

Schrödinger's Silly Cat

Probably the most famous thought experiment in the entire history of physics is that suggested by Erwin Schrödinger in 1935, the same year the EPR thought experiment was proposed. Recall that Schrödinger was one of the old guard and, like Einstein, could not come to terms with the indeterminacy of the quantum world. Consequently, he devised a thought experiment in an attempt to show the follies of the Copenhagen view. Note that this is a thought experiment and no cat was ever forced to be involved in such a procedure!

His experiment goes like this. An atom of a radioactive material is placed in a box along with a Geiger counter which will indicate when the radioactive atom decays. Remember that according to quantum mechanics it is impossible to ever know when an individual radioactive atom will decay, only that it will decay at some stage. Now, apparatus is attached to the Geiger counter which will smash a phial of poison once the counter registers the atom's decay. A cat is placed inside the box, the lid is closed and a certain amount of time allowed to elapse. At any time, the cat is either dead or alive depending on whether the atom has decayed. We do not know which outcome has occurred until we open the box.

In Schrödinger's own words, *the typical feature of these cases is that an indeterminacy is transferred from the atomic to the crude macroscopic level, which can then be decided by direct observation. This prevents us from accepting a "blurred model" so naively as a picture of reality.* (22)

We all know that the cat is dead the moment the atom decays whether we open the box or not and this is Schrödinger's argument against the indeterminate and fuzzy view of reality suggested by the quantum view of nature. He argued that if we accept quantum mechanics and so accept the indeterminacy of the state of the microscopic atom, then we must also accept this indeterminacy when it is transferred to the macroscopic level of the state of the cat which, of course, seems absurd. According to quantum mechanics,

Portrait of Erwin Schrödinger from the book 'Bahnbrecher des Atomzeitalters: grosse Naturforscher von Maxwell bis Heisenberg' by Friedrich Herneck, Berlin: Buchverlag Der Morgen, 1977.

(AIP Emilio Segrè Visual Archives, Brittle Books Collection)

until we open the box and make an observation or measurement, the cat is both indeterminably dead and alive. In a sense the cat is in two superimposed states—a superposition of states, a fogginess or smearing of reality as described by the cat's state function. Opening the box and making an observation or measurement collapses the state function to one of two possible cases as we saw happen with the double slit experiment when we made an observation at one of the slits.

The point Schrödinger was making is that it is quite easy to accept that we do not know when a radioactive atom will decay— that the atom is in a superposition of states, but it is quite another to accept something like a cat to be in a superposition of states. Schrödinger was attempting to highlight what he saw as a flaw in the Copenhagen Interpretation. Common sense tells us that a cat cannot be simultaneously dead and alive but, as we have seen, common sense does not play a big part in quantum mechanics. Common sense is not a useful tool.

One solution to the dilemma is to say the observation is made well before we actually make it. The cat makes the observation and decides whether to be dead or alive but this solution only passes the buck to another player. Of course, it is possible to replace the cat with less conscious organisms or even a computer in order to determine when the actual measurement takes place and the state function collapses. There is however no easy solution to this thought experiment.

What about a group of witnesses to the experiment waiting in the next room? We have opened the shoe box and observed the dead cat but we have not yet told anybody else. For those waiting next door, is the cat still dead and alive simultaneously? Perhaps this situation can be explained by saying that for the guests in the next room the state function has not yet collapsed. We, holding the shoe box, are in a superposition of states. But this argument could be carried back one step further and further, in fact an infinite number of times, so getting us no nearer to an explanation. Does this mean that the entire universe is only real because it has been observed? The American physicist John Wheeler (1911 – 2008) actually proposed that idea as one possible solution. Although the Copenhagen Interpretation of quantum theory still stands as the accepted interpretation, the Schrödinger's Cat dilemma did encourage some researchers to look further for a definite solution.

Bell's Inequality

David Bohm (1917 – 1992) worked in many universities in the United States as well as in England. It was while writing a book on the Copenhagen Interpretation of quantum theory that he began to feel that something was not quite right with it. Upon publication of his book and encouraged by Einstein, Bohm wrote two papers in 1952 describing his new interpretation of quantum theory. The Copenhagen approach suggests that the state function is an abstract wave related to probability. In the model used by Bohm, however, the wave is real and guides particles much like an ocean wave pushes a boat. Particles have precise values of position and velocity and the uncertainty principle only conceals this fact by not allowing their accurate measurement.

Bohm used an improved version of Louis de Broglie's pilot wave model to show that other models were quite capable of producing the traditional Copenhagen outcomes of the quantum theory so it is not necessary to give up a precise, rational and objective description at the quantum level (26). The old guard, comprising Wolfgang Pauli and Werner Heisenberg, were unimpressed by Bohm's work. So was Einstein who had initially encouraged him. Bohm's proposal included the need for non-locality—instantaneous influence at a distance and this was never going to satisfy the great man of physics. In the end, it was clear that Bohm's proposed experiment favoured neither the Copenhagen Interpretation nor Einstein's hidden variables view of reality. Both views could explain the outcomes of Bohm's experiment.

When John Bell (1928 – 1990) read the ideas proposed by Bohm, he was shocked, thinking that any possible alternative to Bohr's Copenhagen Interpretation had been dismissed long ago. In 1932, the mathematician John von Neumann (1903 – 1957) attempted to show that it was not possible to construct an alternative deterministic quantum theory using hidden variables. The Copenhagen Interpretation was valid. Everyone thought this was the end of the matter because of von Neumann's high standing in the scientific community. Von Neumann was well known for his famous quote—in mathematics you don't understand things. You just get used to them (27). And what he said was mostly accepted and people got used to it.

Yet David Bohm thought there was an issue with von Neumann's work although it wasn't clear what it was. Hence his motivation to write the two papers in 1952, which Einstein encouraged, attempting to propose an alternative to the Copenhagen Interpretation. It would finally be up to John Bell to show there was a definite flaw in von Neumann's argument.

Bell began with Bohm's version of the original EPR thought experiment where only one property of a particle, spin, was investigated rather than the two, position and momentum, of the original EPR experiment. Spin can either be spin-up or spin-down and if the system of two particles has total spin of zero, then if one particle is found to have spin-up the other must be spin-down. This arrangement, however, failed to distinguish between the two quantum camps.

Bell proposed an alteration to the experiment by changing the relative angle of the spin detectors to each other. Without delving into the reasoning for doing this, this arrangement made all the difference. Even though this was still a thought experiment, Bell showed that it would be possible to distinguish between the Copenhagen view and the Einstein view of hidden variables and local reality by sampling the correlations between particles for various angles between the spin detectors. Bell's Inequality, as it was called, stated that the Einstein view of a hidden variables/local reality model would only yield correlation coefficients within the limited range between - 2 and + 2 whereas the Copenhagen Interpretation could also yield values outside this range (28). A mathematical inequality is simply an equation that has a greater than or less than symbol in place of the usual equal sign, resulting in a range of allowable values.

John Stewart Bell and his wife Mary in Amherst, MA, in the summer of 1990. (Photograph by Kurt Gottfried, courtesy of AIP Emilio Segrè Visual Archives)

Bell based his inequality on several basic assumptions including:

1. Free will implies random measurements.
2. Reality. Reality exists whether we observe it or not. There is an observer-independent reality. Quantum mechanics is incomplete. There are hidden variables which are currently unknown.
3. Locality. Events are local. There is no instantaneous influence at a distance so there is no faster than light travel.

Bell's Inequality implies that for these assumptions to be upheld, the result of the inequality must lie between - 2 and + 2. If so, the Einstein camp is the winner. If the inequality is violated, (that is, the result is outside the range - 2 to + 2), the Copenhagen Interpretation is vindicated and at least one of the basic assumptions must be wrong.

Bell's Inequality needed to be tested experimentally to decide once and for all between the Bohr and Einstein camps. Einstein's view was that reality was local. There was no instantaneous interaction at a distance, reality existed even when not being looked at, and no information could be transmitted faster than the speed of light. In contrast, Bohr's Copenhagen Interpretation claimed reality to be non-local—there was interaction at a distance and this interaction was instantaneous. If values of the correlation coefficient outside the range from - 2 to + 2 were observed, then Bell's Inequality would be violated and Bohr's Copenhagen Interpretation would be the winner. In 1964, Bell encouraged experimental physicists to attempt a test of his inequality to finally get a verdict.

In 1972, at the age of 30, the physicist John Clauser (born 1942) together with the even younger Stuart Freeman (1944 – 2012) tested the inequality in the laboratory using photons instead of electrons as they found photons to be more suitable to work with. After many hours of data collection, it was found that the inequality was violated. Values of the inequality outside the - 2 to + 2 range were found, so upholding Bohr's Copenhagen Interpretation. It did not take long, however, for the challenges to the methods used to flood in.

In 1983, it was Alain Aspect (born 1947) who mounted the next serious attempt to test Bell's Inequality for his PhD (with Bell as one of his examiners) and again he found the inequality was violated and again there were loopholes in the method that needed closing.

Even so, there was little doubt that the inequality had been violated and further refinements to the experimental technique in more recent years have done nothing to change the outcome. The Copenhagen Interpretation is always vindicated. There is no objective reality. At least one of the three basic assumptions listed above has to give. Typically, people say that the hidden variables assumption has to give, so implying there are no hidden variables in quantum physics.

As Einstein aged, his dominance of scientific thought faded. Niels Bohr's influence on the other hand flourished so much so that followers of his Copenhagen Interpretation attempted to stifle any voice that questioned it. Quantum mechanics seemed to be now entering the realms of philosophy and many physicists drifted away from the central seemingly unexplainable core of the discipline, happy to simply use the outstandingly accurate predictions of the theory when it came to issues of atomic structure and related theories. Quantum mechanics is one of the most successful of scientific theories but, even so, there were those who could not stop tinkering with its fundamental structure.

Other Interpretations

In 1957, Hugh Everett (1930 – 1982) at Princeton University and studying under John Wheeler (1911 – 2008), was encouraged by Wheeler to pursue this problem with the collapsing state function. In fact, what Everett came up with was a non-collapsing view. Everett's proposal is called the Many Worlds Interpretation because it suggests the state function doesn't collapse to just one possible outcome upon us making an observation at the quantum level but, rather, there is no collapse and all possible outcomes continue to co-exist. We only get to see one outcome because the alternatives now exist in other universes. Since all possible outcomes still exist, there is no need to collapse the state function into one outcome.

Unbeknown to Everett at the time, Erwin Schrödinger had come up with a similar idea years before, again with a desire to remove the need for the state function to collapse. As Schrödinger made clear, there was nothing in his wave equation that demanded that this function should collapse. He thought it absurd that the state or wave function could be controlled by two differing entities—by the wave equation itself as well as an independent observer.

The moment we make a quantum observation we split the entire universe into a number of separate universes each containing an alternative outcome. In the case of the cat in the box, when we open the lid the universe divides, one universe containing us viewing a live cat in a box and the other with ourselves viewing a dead cat. Both universes continue to exist and evolve through time until further quantum decisions are forced on each universe at which time they split again. And so on.

To each version of ourselves it appears we have collapsed the state function and forced nature to make a choice between the alternatives but this is only an illusion. The state function did not collapse. Instead, the universe divided at the moment a quantum choice was forced upon it by our quantum measurement. Ultimately there must be an infinite number of universes and an infinite number of different versions of ourselves but we are unaware of these because it is impossible to travel between universes.

The obvious objection to this view of nature is that it is extravagant and extremely wasteful of universe space. Considering the number of quantum events that continually occur not only here on earth but across the entire universe, a mind-boggling array of universes must soon spring up. It seems to be a bad use of space. Perhaps our difficulty in accepting the Many Worlds Interpretation says more about our view of consciousness. Why can't we simultaneously be aware of all superpositions? What is wrong with us? Is our consciousness really that limited? It certainly seems to be.

Everett went very deeply into the matter producing a very complete mathematical description of the many worlds view. Science fiction writers quickly picked up on the idea, creating fictitious parallel world scenarios and, even today, there is still some enthusiasm amongst scientists for the theory.

Notwithstanding Everett's thorough treatise, it is generally conceded however that the many worlds view creates more problems than it solves. Some argued that because we do not observe the universe splitting then the idea cannot be a description of reality. Of course, the actual splitting of the universe implies it cannot be observed simply through the definition of what a universe is. A universe encompasses everything in our experience so there cannot be communication across universes. Everett himself says it best—*the theory itself predicts that our experience will be what in fact it is.* (23)

In 1986, physicist John Cramer (born 1934) produced his Transactional Interpretation of quantum mechanics based on work by John Wheeler and Richard Feynman who developed the time-symmetric or absorber theory. Way back in 1926, Max Born made the suggestion that there was the possibility of paired Schrödinger equations—one describing an advanced wave travelling backwards in time and a retarded wave travelling forward. The transaction is made when these waves interact. As it turns out, the mathematics ends up being the same as that in the Copenhagen Interpretation but the Copenhagen Interpretation applies time in the classical Newtonian way whereas the Transactional Interpretation uses relativistic time. Interestingly, if the speed of light were infinite, the locality issues displayed in Bell's Inequality would be eliminated—local and non-local would be the same, and the Schrödinger equation would be the fully correct description. Cramer's work demonstrates the possibility of a link between relativity and quantum mechanics. (32)

And what next? The mind boggles at the thought of future interpretations. If the past is any indication, then the future will be a fertile field and quantum theory will continue to develop—probably in very mysterious ways. Amidst all the questioning, self-doubting and analysis, Richard Feynman came forth with the statement—*I think I can safely say that nobody understands quantum mechanics . . .Nobody knows how it can be like that.*

CHAPTER 7

CONCLUSION

From its pioneering days in the 1920s, the quantum theory went on to describe the fundamental particles that make up the atom including protons, neutrons and electrons as well as an even more fundamental particle from which both protons and neutrons are built—the quark. After QED came QCD or quantum chromodynamics which described the properties and types of quarks by relating them by metaphor to the more familiar notions of colour and flavour. The metaphors gradually grew more extreme. The collection of fundamental particles and their interactions became known as the standard model of the atom and the quantum mathematics underlying their form was exceedingly complex. As Richard Feynman noted in 1981—*QCD field theory with six flavours of quarks with three colours, each represented by a Dirac spinor of four components, and with eight four-vector gluons, is a quantum theory of amplitudes for configurations each of which is 104 numbers at each point in space and time. To visualize all this qualitatively is too difficult.* (24)

Although some implications of the theory appear quite bizarre, there can be no question that the quantum theory works. It has brought us such inventions as computer chips, computer and television monitors, quantum computing and lasers, as well as explaining all chemical and nuclear reactions. It is probably the most powerful and complete theory in all of science. It describes both the microscopic world of the atom as well as our more familiar macroscopic world. The theory's deep inner mysteries and implications have not in any way hindered the proliferation of practical applications. That a single particle such as an electron is described as being diffusively spread out across a large region of space has ultimately no consequence in the accurate firing of many electrons in a particle accelerator or in an old-fashion cathode ray television picture tube from the 1930s to produce sharp images. Most electrons arrive at the screen at precisely the correct place at precisely the correct moment that probability theory dictates, giving us the sharp images we expect.

Ultimately there is the question of reality. For all its practical uses, the question still left unanswered is whether the theory provides us with an accurate description of nature. After all, this is the prime concern of any scientific inquiry. Does nature really behave in this strange unpredictable way at the microscopic quantum level while behaving quite predictably or deterministically at the macro-level, or are there still some hidden underlying variables yet to be discovered which will reduce the theory to one with which we are more comfortable and one more classically or deterministically based? Quantum mechanics currently says there are no hidden variables. At what point does the microscopic become macroscopic? The questioning continues. Figure 26 shows the domains of the various branches of physics as they currently are.

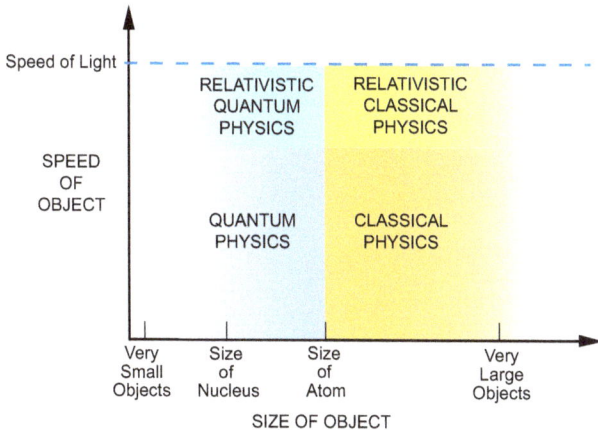

FIGURE 26: Regions in an object speed versus object size graphical space covered by the various domains of physics theories. Can there ever be one all-encompassing treatise?

141

As it stands, quantum theory demands that nature be built on quantum fuzziness and the view that the fundamental proposals of quantum mechanics do provide a complete description of the behaviour of the universe is widespread. Yet, even though it is proclaimed to be the most successful of scientific theories, not everyone can agree on its fundamental assumptions. The discussion continues to this day.

There seems to be an inherent uncertainty in nature. Uncertainty seems to be a fundamental and certain thing, but there is no certainty that this view will always persist. Many alternatives to the Copenhagen Interpretation have been proposed. A major problem with the Copenhagen Interpretation is that it requires an observer to make measurements before the state function, initially in a superposition of possible states, collapses into a single observable one. The mathematics of the quantum theory does not include any mention of an observer. Without that observer, the universe should remain in a superposition of possible states. It should never be created. Work that one out!

APPENDIX 1

A Very Brief Summary of Quantum Physics

New theories only need to be developed when the currently accepted model will no longer predict some new discovery or behaviour. This is what happened with the development of the theory for light. Initially, Newton's particle theory described everything. Then diffraction was discovered and it was found that a wave theory could explain this together with all the previously discovered properties of light. Newton's old particle theory of the 1600s was thrown out but now, as the twentieth century was dawning, Max Planck and Albert Einstein were suggesting once again that a particle theory of light would be required to describe a new aspect of light. But they were suggesting a particle model far removed and far more advanced than Newton's corpuscular theory. Both the wave and particle theories co-exist today because they are both needed to describe all the properties of light. The wave theory is better at explaining interference, as demonstrated by the double slit experiment, while the particle theory has become better at explaining some of the quantum aspects of light.

The double-slit experiment demonstrated the interference of light waves. In this experiment, two light waves interacted to produce a light and dark pattern on a screen when only a uniform and brighter glow across the screen was predicted when two slits are used instead of one.

The quantum theory developed from a discovery that the values of some things in classical physics which were considered continuous, or had a continuum or smooth range of values, were in fact quantized. This simply means that the values of certain commodities had to be specific values and could not be just any value. In 1900, Planck suggested that atoms could only emit light energy in packets of a certain specific size which he termed quanta.

In 1905, Einstein proposed a reversion to the particle theory of light to explain the photoelectric effect. Einstein's idea was the mere beginning of a new particle theory of light which would eventually develop into the full quantum theory and explain things the wave theory of light could never hope to.

Adopting Planck's approach very neatly cleared up several unexplained properties of light emission including what had become known as the ultra-violet catastrophe. Planck declared that the light emitted from atoms came in clearly defined amounts. He did not say that all of light was composed of packets of energy or quanta. This is the step that Einstein took. Planck had discovered the quantization of light but did not realize his idea could lead to a modern-day particle description of light. The equation that Planck had applied to atoms to predict the type of light they would emit, namely $E = hf$ where h is Planck's constant, Einstein now applied to light itself, declaring that light is not a continuous wave as had been thought for so long but rather consists of packets, particles or quanta. We now call these light quanta photons.

The evolving quantum theory inevitably lead to an investigation of the structure of the atom. J. J. Thomson had discovered the electron in 1897 and Ernest Rutherford proposed a model for the atom incorporating a heavy central nucleus surrounded by orbiting electrons much like our solar system. The model however did not account for the fact that, in such a system, the electrons would slowly loose energy and spiral into the nucleus.

Niels Bohr's suggestion was that the electrons did indeed orbit the nucleus like planets but did not gradually lose energy and spiral in because the electrons were forbidden to lose energy gradually. Under his new quantum theory of atomic structure, electrons were only allowed to emit energy in certain quantities. In other words, in whole quanta and not any amount of energy.

What Bohr suggested ran counter to the doctrine of classical physics where quantities such as energy can adopt any value and can change smoothly or continuously from one value to another. Bohr's model combined certain classical concepts such as orbiting electrons behaving like planets while at the same time incorporating some of the more modern quantum ideas of fixed values or quanta. No matter how bad was the structure of Bohr's model, it did work. It explained what Planck had observed many years before—that atoms released light in discrete quantities at certain frequencies. Bohr's model explained how this could happen. If an electron jumped down from one energy state to a lower one, the energy of the light emitted would exactly equal the energy difference between these two levels. Different substances emitted light photons of different colour because the electron energy levels were different distances apart, and different energies meant different frequencies according to $E = hf$. Likewise, electrons could absorb an incoming photon's energy, propelling it to a higher energy level.

The Bohr model went on to explain much about chemical reactions and the periodic table of the elements. It did a superb job of explaining the whole structure of chemistry and chemical reactions, and later the structure of the atomic nucleus. These findings are still used today. Quantum theory is a very successful and practical theory. Much work continued in this field and does so to this day, expanding and filling out the model of the atomic structure.

But there were those theoretical researchers who were a little concerned that there was not a coherent theory underlying all of the findings of the quantum theory. What was now needed was a thorough mathematical treatment encompassing all the facets of the Bohr atomic model. Werner Heisenberg ultimately set the ball rolling. His new mathematical treatise came to be known as quantum mechanics in order to imply a contrast with the classical mechanics view. This new theory was in fact so successful that quantum mechanics replaced classical mechanics, being valid equally for both the microscopic world of the atom and macroscopic world around us.

In 1923, Prince Louis de Broglie took the dual nature of light, being either a wave or a particle, a large step further, suggesting that if a wave-like entity such as light could sometimes behave like a stream of particles with momentum then perhaps, conversely, what we call real particles or objects could sometimes behave like waves. De Broglie suggested that all the equations which had been developed so far to describe the energy, frequency and momentum of a photon of light could equally be applied to any particle. Any object could be ascribed a wave frequency. Quantum mechanics tells us that other particles have wave properties and in fact the heavier the object and therefore the greater the momentum, the smaller the wavelength of that object.

Tennis balls do not behave like waves when projected towards double slits because their associated wavelength is much smaller than the width of each slit. As such, very little diffraction occurs, and tennis balls are seen to behave like particles. Essentially, they go straight through the slits and impact the screen in very predictable straight lines.

In 1925, Heisenberg developed his Uncertainty Principle. He noticed that the quantum equations implied an uncertainty in the parameters. When we confine the frequency of an object to just one value the particle has no position at all—it is spread throughout the entire universe. Sensing the significance of this, Heisenberg seized the moment.

This, he realized, was the connection between the wave and the particle dualism of matter that had eluded everyone. Occupying a small region was a particle-like characteristic whilst the associated frequency range was obviously related to waves. The principle of complementarity precluded the simultaneous use of both the wave and particle characters of matter and now Heisenberg saw a way of quantifying an inverse relationship—less of one character implied more of the other or, in other words, a smaller ambiguity in the position implied greater ambiguity in the frequency. Importantly, this uncertainty in values was an intrinsic property of nature and had nothing to do with the technique or quality of the measuring instruments used.

It was the wave aspect of matter developed by de Broglie that Erwin Schrödinger, in 1926, picked up and carried into his own theory which became known as wave mechanics to distinguish it from Heisenberg's matrix mechanics and Paul Dirac's quantum algebra.

Schrödinger thought he had eliminated the need for quantum steps between the electron energy levels in the atom. He was hoping his theory would show a continuous and smooth increase in possible energy levels for the electron and thought his mathematics could explain what was happening during the quantum leap from one level to another. Schrödinger extended his theory and ultimately showed to his own disappointment that his theory did not preclude the need for quantum leaps between energy levels at all.

Then, around 1926, Bohr tied together all the various quantum findings into what became known as the Copenhagen Interpretation of quantum mechanics. 1926 was also the year Einstein muttered his famous line that *(God) does not play dice.*

Try as he might, Einstein however could not find fault with the quantum theory which he, in fact, had taken part in creating. For the rest of his life, he was uneasy with the statistical or probabilistic character of nature upon which the quantum theory was based. Surely there must be some underlying explanation not yet discovered which would explain everything. Search as Einstein did, none was ever found.

More recently, it is the assumptions of the well-established Quantum Entanglement concept that are receiving attention. Entanglement involves the idea that particles can somehow be linked even though they may be separated by some distance. There is the possibility that information about the state of one can be instantly transferred to the other distant particle. This in turn brings in the issue of instantaneous influence at a distance and hence the need to travel at rates faster than the speed of light. This would contravene Einstein's well-respected and proven relativity theory.

As it stands, quantum theory demands that nature be built on quantum fuzziness and the view that the fundamental proposals of quantum mechanics do provide a complete description of the behaviour of the universe is widespread. Yet, even though it is proclaimed to be the most successful of scientific theories, not everyone can agree on its fundamental assumptions. The discussion is on-going.

APPENDIX 2

Particle Properties of the Photon

Light is often thought of as a wave but if the photon theory of light is considered, then a particle model is more appropriate. This is developed here. From relativity theory, the mass of an object travelling with a velocity v is given by

$$m = \frac{m_0}{\sqrt{1 - \frac{v^2}{c^2}}} \quad \text{(Equation 1)}$$

where m_0 is the stationary or rest mass and c is the speed of light.

Begin to re-arrange this equation by first squaring both sides to eliminate the square root, resulting in

$$m^2 = \frac{m_0{}^2}{1 - \frac{v^2}{c^2}}$$

Multiplying both sides by $1 - \frac{v^2}{c^2}$ gives

$m^2(1 - \frac{v^2}{c^2}) = m_0{}^2$ and, after expanding,

$$m^2 - m^2\frac{v^2}{c^2} = m_0{}^2$$

Multiplying both sides by c^4

$$m^2c^4 - m^2v^2c^2 = m_0{}^2c^4 \quad \text{(Equation 2)}$$

This is true for any object. If the object is a photon, then $v = c$, so

$m^2c^4 - m^2c^4 = m_0{}^2c^4$ and so

$0 = m_0{}^2c^4$ and so

$m_0 = 0$

In other words, the photon has zero rest mass. Conversely, it can be shown that for any particle to travel at the speed of light, its rest mass must be zero. So ordinary particles (which do have a finite rest mass) can never travel at the speed of light.

Substituting $v = c$ into Equation 1 yields $m_0 / 0$ which is infinite. In other words, the mass of any object becomes infinite if it travels at the speed of light.

Since $E = mc^2$, the energy of the object also becomes infinite which is an impossibility. On the other hand, if m_0 is zero, the equation becomes $0 / 0$ which is an allowable but indeterminant quantity.

Also, since $E = mc^2$ and momentum $p = mv$ for any particle, substituting these in Equation 2 results in

$$E^2 - p^2c^2 = m_0^2 c^4$$

For the photon, we know the rest mass $m_0 = 0$, so

$$E^2 - p^2c^2 = 0$$

or, after re-arranging and taking the square root,

$$E = pc$$

so

$$p = \frac{E}{c}$$

This equation is used in Appendix 3.

APPENDIX 3

Wave Properties of the Electron

According to the example depicted in Figure 14, the circumference of the electron orbit is exactly equal to 8 wavelengths, so

$$2\pi r = 8\lambda \quad \text{or, more generally,} \quad 2\pi r = n\lambda \quad \text{(Equation 1)}$$

where n can equal any whole number 1, 2, 3 . . .

Now, for any form of wave motion, it is possible to relate frequency with wavelength via the simple equation

$$\lambda = \frac{v}{f}$$

where λ is the wavelength, f is the frequency and v is the wave velocity. For light, this equation becomes

$$\lambda = \frac{c}{f}$$

where c is the velocity of light. So, λ can now be replaced in Equation 1 above by c/f, giving

$$2\pi r = n\frac{c}{f} \quad \text{(Equation 2)}$$

We have already seen in Chapter 5 that the relationship between momentum p and frequency f (as developed by de Broglie by combining Einstein's two equations $p = E/c$ and $E = hf$) is

$$p = \frac{hf}{c}$$

where p is the momentum of the particle and h is again Planck's famous constant. This equation can be simply re-arranged to give

$$\frac{c}{f} = \frac{h}{p}$$

So, c/f in Equation 2 can be replaced by h/p, giving

$$2\pi r = n\frac{h}{p}$$

And this can be re-arranged to give

$$pr = n(\frac{h}{2\pi})$$

which is the same as the series of numbers obtained by Niels Bohr (0, $h/2\pi$, $2h/2\pi$ . ..) for the angular momentum which is represented here by pr.

Note also that this equation has a shape very similar to that of the Uncertainty Principle derived by Werner Heisenberg and discussed in Chapter 5.

APPENDIX 4

The Square Root of Minus One

Finding a value for the square root of -1 represents a dilemma. It seems that all possible real numbers are already used up. The simplest way to begin to work out a solution is to try squaring, rather than taking the square root, of various numbers. Squaring a number means multiplying it by itself. Taking the square root of a number means finding a number when multiplied by itself gives you the original number. So, 1 squared is 1. Working back from this, the square root of 1 must also be one.

Let's try another. The square of -1 is also 1 since two negative numbers multiplied together make a positive number. So, the square root of 1 can either be +1 or -1 because both of these numbers when squared give 1 as the result. Summarising, we have

$$1^2 = (-1)^2 = 1$$

and

$$\sqrt{1} = 1 \text{ or } -1.$$

What we now need to find is a number which when squared gives us -1. If we find this number it will be the answer to our question—the square root of minus one. From the above discussion it should be clear that there is no real number because no number when multiplied by itself gives a negative answer.

The solution is to simply define a symbol, namely i, to represent the square root of -1. i cannot be considered a real number like 1, 2, 3 . . because as we saw, no real number can be found to represent the square root of minus 1. i is termed an imaginary number to distinguish it from the real numbers. This, in hindsight, is a poor choice of title because imaginary numbers are just as useful as real numbers and are used to perform many duties in mathematical calculations.

NOTES

This list of notes, referenced by a number in the text, identifies specific sections in other texts which deal with the current discussion in more detail.

1. White, M. and Gribbon J. (1994). *Einstein — a Life in Science*. Simon and Schuster Ltd, London. (Page 213)

2. Gribbon, J. (1985). *In Search of Schrödinger's Cat*. Corgi Books, Great Britain. (Page 77)

3. Bronowski, J. (1973). *The Ascent of Man*. British Broadcasting Corporation, London. (Page 360)

4. Hart, M. (1993). *The 100 — A Ranking of the Most Influential Persons in History*. Simon and Schuster Ltd, London. (Page 293)

5. Hart, M. (1993). *The 100 — A Ranking of the Most Influential Persons in History*. Simon and Schuster Ltd, London. (Page 291)

6. Gribbon, J. (1985). *In Search of Schrödinger's Cat*. Corgi Books, Great Britain. (Page 84)

7. Penrose, R. (1989). *The Emperor's New Mind*. Vintage, London. (Page 295)

8. Resnick, R. and Halliday, D. (1966). *Physics*. John Wiley and Sons, Inc., New York. (Page 589)

9. Penrose, R. (1989). *The Emperor's New Mind*. Vintage, London. (Page 299)

10. Gribbon, J. (1985). *In Search of Schrödinger's Cat*. Corgi Books, Great Britain. (Page 107)

11. Penrose, R. (1989). *The Emperor's New Mind*. Vintage, London. (Page 227)

12. Gribbon, J. (1985). *In Search of Schrödinger's Cat*. Corgi Books, Great Britain. (Page 110)

13. Penrose, R. (1989). *The Emperor's New Mind*. Vintage, London. (Page 299)

14. Penrose, R. (1989). *The Emperor's New Mind*. Vintage, London. (Page 314)

15. Hart, M. (1993). *The 100 — A Ranking of the Most Influential Persons in History*. Simon and Schuster Ltd, London. (Page 236)

16. Penrose, R. (1989). *The Emperor's New Mind*. Vintage, London. (Page 361)

17. Gribbon, J. (1985). *In Search of Schrödinger's Cat*. Corgi Books, Great Britain. (Page 117)

18. Gleick, J. (1992). *Genius — Richard Feynman and Modern Physics*. Little, Brown and Company, Great Britain. (Page 378)

19. Feynman, R. P. (1965). *The Character of Physical Law*. Penguin Books Ltd., England. (Page 145)

20. Gleick, J. (1992). *Genius — Richard Feynman and Modern Physics*. Little, Brown and Company, Great Britain. (Page 139)

21. Price, H. (1996) *Time's Arrow and Archimedes' Point*. Oxford University Press, Oxford. (Page 202)

22. Price, H. (1996) *Time's Arrow and Archimedes' Point*. Oxford University Press, Oxford. (Page 203)

23. Price, H. (1996) *Time's Arrow and Archimedes' Point*. Oxford University Press, Oxford. (Page 199)

24. Gribbon, J. (1985). *In Search of Schrödinger's Cat*. Corgi Books, Great Britain. (Page 242)

25. Gleick, J. (1992). *Genius — Richard Feynman and Modern Physics*. Little, Brown and Company, Great Britain. (Page 403)

26. Kumar, M (2014). *Quantum*. Icon Books Ltd., London. (Page 335)

27. Wikipedia (*John von Neumann*)

28. Kumar, M (2014). *Quantum*. Icon Books Ltd., London. (Page 346)

29. Gribbon, J. (1989). *Quantum Rules, OK!* Inside Science, New Scientist, September 1989, IPC Magazines, London.

30. Cornwell, John (2003). *Hitler's Scientists—Science, War and the Devil's Pact*. Penguin Books Ltd, London. (Page 170)

31. Cornwell, John (2003). *Hitler's Scientists—Science, War and the Devil's Pact*. Penguin Books Ltd, London. (Page 318)

32. Gribbin, J (2019) *Six Impossible Things—The "Quanta of Solace" and the Mysteries of the Subatomic World*. Icon Books Ltd., London. (Page 101)

33. Resnick, R and Halliday, D (1966). *Physics*. John Wiley and Sons, Inc., New York. (Page 1212)

REFERENCES

If the reader wishes to delve further into the topics discussed in this book, then the following list will be extremely useful. These books explain in more detail and expand on the many issues and ideas discussed in this title.

Baker, J (2007). *Fifty Physics Ideas*. Quercus Publishing, London.

Bronowski, J (1973). *The Ascent of Man*. British Broadcasting Corporation, London.

Cornwell, John (2003). *Hitler's Scientists—Science, War and the Devil's Pact*. Penguin Books Ltd, London.

Feynman, R P (1965). *The Character of Physical Law*. Penguin Books Ltd, England.

Feynman, R P (1985). *QED—The Strange Theory of Light and Matter*. Penguin Books Ltd, London.

Gleick, J (1992). *Genius—Richard Feynman and Modern Physics*. Little, Brown and Company, Great Britain.

Gribbin, J (1985). *In Search of Schrödinger's Cat*. Corgi Books, Great Britain.

Gribbin, j (1999). *Q is for Quantum—Particle Physics from A to Z*. Phoenix, Orion Books Ltd., London.

Gribbin, J (2019) *Six Impossible Things—The "Quanta of Solace" and the Mysteries of the Subatomic World*. Icon Books Ltd., London.

Hart, M (1993). *The 100—A Ranking of the Most Influential Persons in History*. Simon and Schuster Ltd, London.

Kumar, M (2014). *Quantum*. Icon Books Ltd., London.

Ohanian, H. C. (2008). *Einstein's Mistakes—The Human Failings of Genius*. W. W. Norton and Co., New York, London.

Penrose, R (1989). *The Emperor's New Mind*. Vintage, London.

Price, H (1996) *Time's Arrow and Archimedes' Point*. Oxford University Press, Oxford.

Rankin, R (2011). *Einstein's Relativity*. Rankin Publishers, Brisbane.

REFERENCES

Resnick, R and Halliday, D (1966). *Physics*. John Wiley and Sons, Inc., New York.

White, M. and Gribbon J. (1994). *Einstein—A Life in Science*. Simon and Schuster Ltd, London.

INDEX

of people and concepts for Chapters 1 to 7 only. Bold type indicates the item is in a picture or diagram.

INDEX

www.ingramcontent.com/pod-product-compliance
Lightning Source LLC
Chambersburg PA
CBHW040858210326
41597CB00029B/4887